3D Concrete Printing

SCIENCES

Mechanics, Field Director – Gilles Pijaudier-Cabot

Civil Engineering and Construction, Subject Head – Fabrice Gatuingt

3D Concrete Printing

State of the Art and Applications

Coordinated by
Arnaud Perrot
Yohan Jacquet

iSTE

WILEY

First published 2025 in Great Britain and the United States by ISTE Ltd and John Wiley & Sons, Inc.

ISTE Ltd
27-37 St George's Road
London SW19 4EU
UK

www.iste.co.uk

John Wiley & Sons, Inc.
111 River Street
Hoboken, NJ 07030
USA

www.wiley.com

Library of Congress Control Number: 2024946962

British Library Cataloguing-in-Publication Data
A CIP record for this book is available from the British Library
ISBN 978-1-78945-200-6

ERC code:
PE8 Products and Processes Engineering
PE8_3 Civil engineering, architecture, offshore construction, lightweight construction, geotechnics

Contents

Preface

Arnaud PERROT
IRDL, Université de Bretagne Sud, Lorient, France

P.1. Purpose of the book

The rapid emergence of 3D concrete printing technology has opened up exciting new perspectives in construction and engineering. Because of this rapid dynamic, we must now understand this craze and explore the limits and potential of this innovation in the field of construction. This can be done by taking stock of developments in printing technologies and applications.

Additive manufacturing of concrete structures offers significant advantages such as cost reduction, customization of architectural designs, speed of execution, improved safety and working conditions, and minimization of the volume of material used (structural efficiency, multi-functionality) and waste.

Since the early 2000s, developments and projects relating to 3D printing have been multiplying at a steady pace, resulting in an exponential increase in the number of scientific studies on the subject. The subject, which initially focused almost exclusively on the formulation of materials and robotics, now encompasses many other aspects, such as quality control, applications, the regulations that are beginning to emerge (ISO 2023), mechanical operation and characterization, durability and environmental impact, etc.

In 2019, I edited a first book on the subject of concrete 3D printing (Perrot 2019), which is now becoming almost obsolete due to recent advances in this field. Thus, writing a new, updated book dedicated to 3D concrete printing has become

3D Concrete Printing,
coordinated by Arnaud PERROT and Yohan JACQUET. © ISTE Ltd 2025.

essential in order to document, analyze and propagate current knowledge related to this revolution in the field of construction. This book could serve as a comprehensive guide, covering the technical aspects, current challenges, potential applications, as well as the difficulties and limitations of printing methods. It would thus provide an indispensable resource for researchers, students of civil engineering, materials engineering and robotics, and construction professionals looking to stay at the forefront of this major development in the industry.

To write this book, a multidisciplinary team of academics and industrialists was brought together to deal with the various aspects of 3D printing of cementitious materials such as concrete.

P.2. Structure of the book

This book is structured into eight chapters, each dealing with a specific aspect of 3D concrete printing. In the first part, an overall presentation of concrete printing processes (which is ultimately a family of very multifaceted digital processes) is given, with a summary of the existing classifications proposed by working groups or research teams.

The next two chapters focus on the extrusion deposition printing technique. Chapter 2 depicts this process, and Chapter 3 describes the characterization techniques developed to mix-design printable materials on a laboratory scale, characterize their mechanical behavior once hardened and organize quality control during production in an industrial context.

Chapter 4 explains other families of printing and digital manufacturing techniques that have been developed for cementitious materials.

Chapter 5 defines the mechanical behavior of printed materials and structures, as well as the structural design of printed concrete. To take the structural use of printed concrete a step further, Chapter 6 presents the reinforcement techniques that have been and are currently being developed in order to guarantee the necessary mechanical performance.

Chapter 7 outlines the numerical simulation tools used to simulate the 3D concrete printing process, showing the specific characteristics of the material and its behavior that need to be taken into account in the additive manufacturing of cementitious materials.

Finally, Chapter 8 focuses on the environmental analysis of printed structures. It highlights the potential benefits of 3D printing as part of an eco-construction approach and identifies the areas with the greatest impact in terms of carbon emissions.

November 2024

P.3. References

ISO (2023). Additive manufacturing for construction: Qualification principles – Structural and infrastructure elements. Standard, ISO/ASTM 52939.

Perrot, A. (ed.) (2019). *3D Printing of Concrete: State of the Art and Challenges of the Digital Construction Revolution*. ISTE Ltd, London, and John Wiley & Sons, New York.

1

3D Concrete Printing: Technologies, Applications and Classifications

Arnaud PERROT[1], Yohan JACQUET[2] and Sofiane AMZIANE[3]

[1] *IRDL, Université de Bretagne Sud, Lorient, France*

[2] *Technische Universität Berlin, Germany*

[3] *Institut Pascal, Université Clermont-Auvergne, Clermont-Ferrand, France*

1.1. Introduction: the different facets of printing

Since its initial development in the early 2000s, 3D printing has always been a polymorphic process, depending on the type of material used and the robotic system employed. Initial developments such as Prof. Koshnevis' Contour Crafting concept (Khoshnevis 2004; Khoshnevis et al. 2006; Buswell et al. 2018), based on an extrusion system, or Enrico Dini's D-Shape process, based on a selective particle bonding printing system (Colla et al. 2013), illustrate this diversity, which has continued to increase ever since. 3D concrete printing has almost as many variations as there are printers and companies working on the subject.

The multifaceted nature of printing is reflected in ASTM F2792-12A (ASTM International 2012), which defines seven categories of additive manufacturing processes that can be used for any type of material. These categories can also be related to the different processes used today for concrete and cementitious materials.

Although there is a wide range of possible processes, 3D concrete printing techniques (which often involve mortar more frequently than concrete, as the

3D Concrete Printing,
coordinated by Arnaud PERROT and Yohan JACQUET.

presence of gravel requires large pumps, print heads and robots) share many points in common, particularly in the use of digital technology and robotics (Perrot and Amziane 2019).

The integration of robotics and digital technology means that the starting point is the design of a 3D model of the object to be printed, the drafting of instructions that the robot can understand and finally the production of the object by the robot.

In this chapter, we will start by presenting these stages, which are common to the printing process in general. We will then present the current applications of the 3D concrete and mortar printing, focusing on the materials used and alternatives to conventional cementitious materials.

Finally, we will present several classifications that help position a given printing process in the world of digital concrete manufacturing.

1.2. 3D printing: from digital model to physical object

1.2.1. *From digital model to print*

The first step in 3D printing is to design the part to be printed. This design stage always begins with creating a 3D model in CAD (computer-aided design) software.

Many 3D design solutions exist, but they differ according to the philosophy of the tool used and the complexity of the object to be modeled.

For modeling simple, perfectly vertical shapes, we may draw the deposition contours of a layer to be printed on the deposition plane and then extrude this surface in the direction perpendicular to the deposition plane (generally the vertical axis). This design and subsequent printing strategy is often referred to as 2.5D printing.

For more complex shapes with overhangs, it is necessary to work in three dimensions and use modeling solutions to represent lines and surfaces in a 3D reference frame.

Finally, in order to design architectural shapes with patterns, parametric 3D design is an extremely useful tool for obtaining shapes with repetitive patterns that provide added aesthetic value or additional functions (noise absorption, sun shading, etc.). The use of tools, such as Rhino/Grasshoppers, is among the most popular parametric design solution (Figure 1.1). This solution uses a graphical programming system to manipulate shapes and objects and repeat actions to create 3D patterns.

As part of a BIM project, geometric design in software compatible with the IFC format, the BIM exchange format may be of interest. The 3D model then needs to be exported in a format compatible with the software used to generate the robot trajectory. The goal is to export the file in a format that can generate a G-Code (Geometric Code), which is a file containing a sequence of instructions that drives a 3D printer. The .STL, .obj and .amf formats can be used for this purpose.

These export files are then imported into a "slicer", which is used to program the trajectory of the print head and set several parameters, such as the start of extrusion, the position of the print head during deposition, the infill of a volume (pattern and density) or the start position of the print. The defined robot trajectory must then be optimized to ensure no collisions during printing and good contact between adjacent or superimposed deposits.

It is also important to note that it may be necessary to convert the G-Code to a language compatible with the printing system (such as six-axis robots).

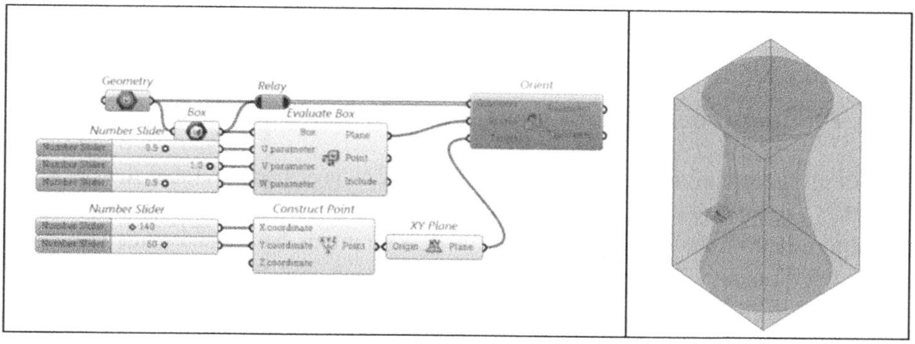

Figure 1.1. *Parametric design – programming and printed parts. For a color version of this figure, see www.iste.co.uk/perrot/3dconcrete.zip*

1.2.2. *Printing processes*

ASTM F2792-12A (ASTM International 2012) defines seven different categories of additive manufacturing methods. This standard has become the benchmark for defining 3D printing terminology.

Among these categories, the material extrusion method is certainly the most widely used, particularly in polymer 3D printing by melt extrusion/deposition. Plastic 3D printing that involves fusing filaments of different types of plastic through a thermo-regulated extrusion system has significatively contributed to the democratization of the printing process.

Jetting, either of binders or materials such as polymers or molten metals, is also listed as one of the techniques in the ASTM standard.

The lamination of sheets consists of gluing numerous sheets of material together to form a larger object. The layers can be pre-cut before bonding, or a cutting step can be carried out after bonding the sheets to create a complex shape.

Finally, the last three methods are specific techniques requiring the application of an additional energy source to ensure the fusion or hardening of the material. For example, fusion on a powder bed or the deposition of material under a directed energy flow, widely used for metals, requires the application of a significant amount of energy to melt the material, ensuring the bonding of the material once it has returned to ambient temperature. Finally, photopolymerization uses light to locally solidify a photosensitive resin to bind particles together and form a resistant matrix. This method is often called stereolithography.

1.2.3. *Printing processes for cementitious materials*

Not all the methods listed in ASTM F2792-12A have been adapted to concretes or other cementitious materials, especially given their non-dependence on temperature and light. However, numerous developments have been made (Buswell et al. 2020) using different techniques for different applications and at different scales.

Concrete printing by extrusion deposition was the first technique to be used. The "contour crafting" developed by Khoshnevis et al. (2006) or the "concrete printing" method from Loughborough University (Le et al. 2012; Lim et al. 2012; Perrot 2022) used the extrusion/deposition method before 2010. Since then, the 3D concrete extrusion printing method, often referred to as 3D concrete printing, has undergone many further developments with numerous variations in the printhead, notably with the advent of two-component formulations, the scale of realization, the robotic system used and finally the intended application (Wangler et al. 2016, 2022; Buswell et al. 2018). The deposition extrusion process can also be coupled with other sub-processes to perform other tasks, such as positioning reinforcement, applying a curing agent or improving surface finish (Buswell et al. 2020).

The technique of printing by injection into particle beds has also been adapted to cementitious materials (Lowke et al. 2018, 2020, 2022). In this category, two processes can be distinguished; the first concerns the technique of selective binding of aggregates by localized injection of a cementitious grout (Pierre et al. 2018; Weger et al. 2021) and the second concerns the injection of water to locally activate a powder bed containing a hydraulic binder such as cement (Mai et al. 2022; Zuo et al. 2022). Interestingly, this method was used very early on in Enrico Dini's early work in the first half of the 2000s (Lowke et al. 2018).

The Shotcrete method, on the other hand, is similar to the "Material Jetting" method, in which a cementitious material is projected onto an object (Lindemann et al. 2019; Hack and Kloft 2020).

Finally, other so-called "digital concrete" methods appeared in the scientific literature in 2010, such as smart dynamic casting, which involves automating and robotizing the slipforming process (Lloret-Fritschi et al. 2020; Szabo et al. 2020), and the Mesh Mould technique (Hack and Lauer 2014), in which robots create a permeable formwork that can then be used to pour or apply concrete. Finally, we should also mention optimized formwork printing methods that minimize the amount of material needed to produce complex shapes which can withstand the pressure exerted by freshly poured-in-place concrete (Burger et al. 2020; Lloret-Fritschi et al. 2022).

1.3. 3D concrete printing – application examples

1.3.1. *Prefabrication*

1.3.1.1. *Support-free printing*

The use of 3D concrete printing in the prefabrication of concrete parts, which are then assembled on-site to increase precision and/or production rates (Volpe et al. 2021), or to create specific functions with unique or complex geometries (such as artificial reefs, stormwater basins, voussoirs for a bicycle bridge, prefabricated staircases, etc.; Salet et al. (2018); Ly et al. (2021)) is now commonplace. Among the additive manufacturing methods used for concrete, some do not require supports, such as extrusion-deposition methods, which are particularly developed today.

Numerous companies have developed their own printing solutions and related businesses. Examples include XtreeE in France, Hyperion Robotics in Finland, Weber, TAM and Vertico in the Netherlands, and Sika in Switzerland.

All these companies use six-axis robots to print precast-printed concrete elements. Some of these companies also use a rail on which the robot is placed to increase the available printing volume (Figure 1.2). A system for preparing and transporting the material by pumping is also a standard feature of printing plants.

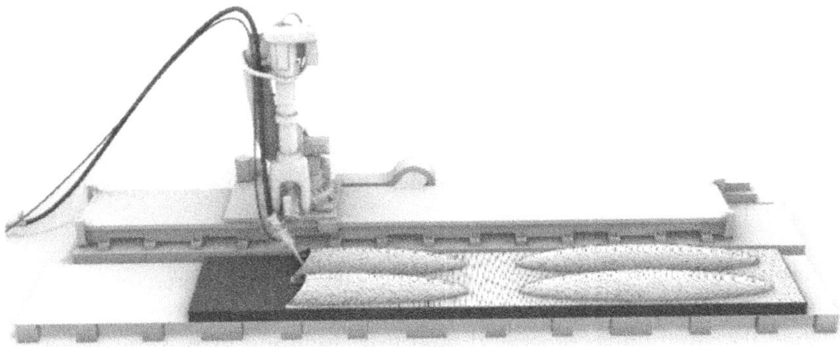

Figure 1.2. *Example of the installation of a 3D printing system consisting of a six-axis robot and a complementary linear conveyor (Lowke et al. 2024). For a color version of this figure, see www.iste.co.uk/perrot/3dconcrete.zip*

Prefabrication is often used to manufacture small parts or parts with a small cross-section. As a result, the length of the printing contour (the length of the filament of a single printed layer) imposes very short deposition times, requiring a rapid rate of elevation of the structure being printed. This fast printing rate induces rapid loading of the structure, which is not compatible with the structuring kinetics of traditional cements (Reiter et al. 2018).

There are two solutions to this problem. The first is to develop single-component formulations, which have a period of rheological stability to enable mixing and transport to the print head, followed by a period of rapid structuring of the material to enable fast printing. The other solution uses two-component materials and consists of injecting a texturizing agent or a setting accelerator at the print head to enable rapid setting kinetic growth as soon as the material is deposited.

Each method has its advantages and disadvantages: the single-component solution does not tolerate interruptions to the printing process, as otherwise, there is a risk of curing in the pump lines and consequent blockage. The two-component solution, on the other hand, adds an in-line mixing stage just before deposition, which requires a careful choice of product to accelerate curing and the design of a more complex print head with an in-line injection and mixing system (active or passive) (Tao et al. 2021; Boscaro et al. 2022; Wangler et al. 2022).

A specific two-component process called dual pumping has been developed by Ghent University (Tao et al. 2022) and enables two fluid materials with equivalent volumes to be pumped, which, once mixed just prior to extrusion, undergo rapid mechanical structuring, enabling high printing speeds.

Robotic complexity also plays an important role in 3D printing (Figure 1.3). A three-axis robot (Cartesian, delta or a six-axis robot using only three-axis) will enable a structure to be printed with a cylindrical filament cross-section, a print head that is always vertical, and limited freedom of form. Switching to four-axis adds rotation around the vertical direction and enables the use of perfectly cylindrical layer cross-sections, which is of interest for obtaining smooth vertical surfaces. Finally, using the robot's six-axis allows printing with changing printhead orientations, providing greater freedom of form and enabling more efficient construction of arches or domes (Carneau et al. 2020).

Figure 1.3. *Type of structures printed according to the type of robot used. (Top right photo: Lowke et al. 2024; Bottom right photo: XtreeE). For a color version of this figure, see www.iste.co.uk/perrot/3dconcrete.zip*

Printed shotcrete also makes it possible to achieve high printing speeds for precast elements, as shown by studies carried out at the Technical University of Braunschweig (Hack and Kloft 2020). 3D printing of shotcrete allows greater freedom of form.

The smart dynamic casting technique also requires no support to produce prefabricated parts such as columns or slender and slender walls (Lloret-Fritschi et al. 2017; Szabo et al. 2020). The use of a six-axis robot fitted with variable-section formwork enables these elements to be produced by adjusting the speed at which the formwork rises to match the speed at which the cementitious material is structured. If the material leaving the sliding formwork is too fluid, the structure collapses, whereas if it is too rigid, the friction at the formwork wall becomes greater than the tensile strength of the printed material section, which becomes brittle and breaks into two parts (Craipeau et al. 2021).

1.3.1.2. *Printing with support*

It is also possible to produce prefabricated parts using 3D printing techniques that use supports. These methods allow greater freedom of form, enabling the creation of parts with cantilevers that transition from a vertical to a horizontal structure.

The particle-bed printing technique can therefore be used to produce parts with complex shapes whose surface roughness depends on the size of the particles used. For example, the selective bonding 3D printing technique will produce parts with a rougher surface than the activation bonding technique.

Similarly, 3D printing by depositing material in a threshold fluid, such as carbopol or limestone suspensions, enables lattice structures to be produced, making it possible to use topological optimization methods for structural dimensioning (De Schutter et al. 2018; Vantyghem et al. 2020) and thus save material for the same mechanical function. It is also possible to print ultra-high-performance concrete in ordinary concrete for the same reasons of mechanical efficiency (Lowke et al. 2021).

Finally, methods for pouring concrete into printed formwork fall into this category. Since the Yhnova house printed in 2017 in Nantes with polyurethane formwork (Furet et al. 2019), numerous advances have been to reduce the weight and thickness of formwork while guaranteeing faster filling speeds, as in the In-crease (Lloret-Fritschi et al. 2022) or Eggschell (Burger et al. 2020) projects at ETH Zurich.

1.3.2. *On-site printing*

Current advances and technologies show that on-site 3D printing is reserved for two types of application: shotcrete and extrusion-deposition techniques. In general, these methods do not use supports except on a case-by-case basis (such as crossing openings with prefabricated lintels) (Tay et al. 2019). They must also be able to perform a structural function in their own right, and therefore incorporate reinforcements (Kloft et al. 2020; Menna et al. 2020).

Shotcrete technology can be seen as an evolution of automated shotcrete wall construction techniques for tunneling and retaining structures (Heidarnezhad and Zhang 2022). These techniques can also be used on-site to build structures independently, without relying on supports such as the soil in tunneling or geotechnical work (Girmscheid and Moser 2001).

When printing on-site, regardless of the method, it is important to bear in mind that the printed material is directly exposed to an uncontrolled environment (wind, rain and sun), unlike the conventional casting process. This direct and prolonged exposure can lead to unexpected changes in the material composition which are potentially deleterious to the material's properties once hardened (Keita et al. 2019) such as lack of water for cement hydration due to excessive drying, or leaching of the structure by heavy rain, etc. This constraint must therefore be considered in the material specifications or by implementing additional precautions such as tenting, covering or applying a curing product.

Extrusion deposition printing is probably the most developed technique for on-site 3D concrete printing. Several robotic systems have been developed to print various sizes (Xiao et al. 2021).

For small constructions, such as single-family homes or small buildings, fixed robots such as the delta robots developed by WASP (Gomaa et al. 2022; Moretti 2023) or the transportable robotic arm developed by Construction 3D (https://www.constructions-3d.com/ n.d.) are ideal.

For larger projects, the use of a large-scale Cartesian printer is a solution that has been developed for printing structures such as collective housing or wind turbine mast elements (Savić et al. 2020). This involves setting up a gantry crane equipped with a print head as an end effector (Ahmed 2023). This solution appears to be the most common on the market today, as demonstrated by the success of printers developed by Cobod.

For even larger scale impressions, it is possible to rely on cable robot technology (Barnett and Gosselin 2015; Izard et al. 2017) or on the adaptation of current construction machinery such as cranes and crane trucks (Gaël et al. 2023). However, there are still technological hurdles to overcome regarding positioning the end effector under the effect of wind, rain and external stresses. It should be noted that the idea of using cable robots as 3D printers dates back some 15 years (Pott et al.

2010), with conclusive trials that have been carried out with raw earth. The structure needed to set up a cable robot seems relatively light and can cover large surfaces. The conversion of construction site equipment into 3D printers remains at the concept stage for now, as seen in the ConPrin3D project developed by the Technical University of Dresden (Schack et al. 2017; Nerella and Mechtcherine 2019) or other initiatives (Gaël et al. 2023). Interestingly, stabilization or compensation techniques to ensure printhead trajectory are currently under development, notably by implementing gyroscopes acting to maintain printhead position.

Finally, recent work is looking at the use of multiple mobile robots for 3D printing structures or structural elements on-site (Zhang et al. 2018; Dörfler et al. 2022). This solution eliminates the need to install a structure such as an overhead crane before printing and opens the way to collaborative work by robots. This collaboration can also be used to share optimal tasks, such as placing reinforcement or lintels simultaneously with printing, either by extrusion or concrete spraying.

More recently, the use of an army of collaborative drones to carry out construction work has been imagined and validated on a small element (Zhang et al. 2022).

However, several challenges remain, such as properly managing robot tracking to prevent robot collisions without hampering production and adapting the locomotion of mobile robots to the topography of the worksite and the load to be carried.

1.3.3. *Toward the democratization of 3D concrete printing?*

Today, 3D concrete printing is experiencing a significant diversification in its applications and scales of realization. This diversity implies a very strong link between the properties of the material in its fresh state (and changes over time, notably the speed of mechanical structuring) (Perrot et al. 2016; Roussel 2018) and the printing process used.

These properties need to be controlled and monitored to guarantee the success of the printing process and ensure a robust production process. This is why quality control methods are currently being developed (Nicolas et al. 2022) to provide a means of controlling printability, similar to the workability tests for ordinary concrete poured into formwork using Abrams cone slump tests (Roussel and Coussot 2005). The RILEM PFC (performance requirements and testing of fresh printable cement-based materials) technical committee is working on the development of these tests.

It is also important to note that concrete printing processes mainly use mortar. However, the on-site printing of larger structures opens up the possibility of printing concrete (Ji et al. 2022), thereby reducing the cost and environmental impact of the printed material.

The printing process itself results in the creation of a structure whose constituent material may exhibit anisotropic behavior as a result of layering. This particularity must be taken into account in the design codes currently being developed by international working groups led by RILEM (TC ADC – Assessment of Additively Manufactured Concrete Materials and Structures), fiB (Fédération Internationale du Béton) and ASTM.

An initial ISO/ASTM standard describing concrete 3D printing was published at the end of 2023 (ISO/ASTM 52939 2023).

The design of the structure must also consider the reinforcement techniques used to give the structure the ability to resist tensile or shear stresses (Kloft et al. 2020; Menna et al. 2020).

These points will be developed in the following chapter of this book.

1.4. Classification of concrete printing processes

3D concrete printing is a polymorphous process that needs to be well described in order to be able to properly design the robotic complexity and the specifications of the printable cementitious material. Similarly, for a given concrete printer, we need to know what type of parts can be printed, on what scale and in what environment. A number of projects have focused on printing process classification systems that can be used to describe the specifications of a process and the possibilities available.

1.4.1. *Classification proposed by Duballet and co-authors*

An initial ranking of concrete printing processes has been proposed by Duballet and co-authors (Duballet et al. 2017). Based on the characteristics of the process and the printed part, the team of authors developed a classification taking into account the complexity of the robotic system, the environment and the geometry of the part to be manufactured. This classification provides an objective description of the resources required to print a given type of object.

For the moment, however, the classification is limited to the extrusion/deposition technique, although it can be transposed or adapted for other methods of printing cementitious materials. To enable the types of printing to be classified, the authors have defined a nomenclature by defining the characteristics of the process.

1.4.1.1. *Scale of the printed object*

The first classification parameter is the scale of the printed object. As we saw earlier, the size of the object to be printed will necessarily have an impact on the type of robot and its workspace. The scale of the object to be printed will also influence the deposition time (length of the contour to be printed) and impact the properties of the material in its fresh state (speed of structuring). The authors of the classification list four typical sizes of objects to be printed:

– x_0^0: object less than 1 m in size;

– x_0^1: object whose size is that of a standard building element: 1–4 m, for example, a beam or a column;

– x_0^2: object the size of a small single-family home: 5–10 m;

– x_0^3: object the size of an entire building.

1.4.1.2. *Scaling the cross-section of printed material*

The second grading parameter is that of the printed material filament. This scale depends on the rheology of the material and the cross-section of the print head, and impacts the surface roughness of the printed structure. The cross-sectional scale of a deposited layer also influences the maximum aggregate size of the cementitious mix and the printing speed. The authors define four different scales:

– x_e^0: layer thickness less than 8 mm;

– x_e^1: layer thickness between 8 mm and 5 cm;

– x_e^2: layer thickness between 5 cm and 30 cm;

– x_e^3: layer thickness greater than 30 cm.

Note that printing time is directly related to the ratio between the size of the object to be printed and the size (height) of the filaments deposited. Print time will therefore be the product of the time required to deposit a layer multiplied by the ratio between the height of the part to be printed and the height of a layer.

1.4.1.3. *Printing environment*

The authors of the ranking highlighted three possibilities for a direct printing environment:

– e^0: direct on-site printing;

– e^1: printing in a small mobile unit on site, allowing hygrometry and temperature control and shelter from the wind;

– e^2: printing in the prefabrication plant.

The environment will have a direct impact on the robustness of the process to be implemented. In a protected atmosphere (second or third solution), the process may be easier to implement than on site, where climatic variations and actions will have to be taken into account (protection of printed material, precision of robot positioning).

1.4.1.4. *Component assembly conditions*

For the printing of prefabricated parts assembled on site, the authors defined four assembly conditions that can occur during the partial (or total) printing of a building:

– a^0: no assembly;

– a^1: assembling several elements to form a larger one;

– a^2: handling elements to place them in their final position;

– a^3: assembling unprinted external elements after printing.

It is important to note that during the construction of a building, several stages require the assembly of elements. For example, printed and unprinted elements (conditions a^2 and a^4) can be assembled on the same construction project.

1.4.1.5. *Use of supports*

3D concrete printing by extrusion/deposition has its limitations in terms of freedom of form, since the stability of the structure must be ensured during printing. It is therefore possible to use temporary or permanent supports, as proposed by the authors of the classification. Four categories of support have thus been listed:

– s^0: no support;

– s^1: printed support left in place;

– s^2: printed support removed at end of printing;

– s^3: external support left in place;

– s^4: external support removed at end of printing.

As with assembly, several solutions can be used simultaneously when printing an object.

1.4.1.6. *Robot complexity*

Several robot configurations are possible for printing 3D structures. Collaboration between several robots to print complex structures or make assemblies can be an effective solution:

– r^0: use of a three-axis robot;

– r^1: use of a six-axis robot;

– r^2: use of a rail-mounted (or mobile) six-axis robot;

– r^3: use of two six-axis collaborative robots;

– r^4: use of two rail-mounted (or mobile) six-axis collaborative robots;

– r^5: use of a six-axis robot mounted on a rail (fixed or mobile);

– r^6: use of a six-axis robot mounted on a three-axis robot;

– r^7: use of two six-axis robots mounted on three-axis robots.

The first five parameters in the ranking will influence the specifications of the right robot for a given print job. Experience and difficulties encountered will determine the minimum robotic complexity.

1.4.1.7. *Example of classification and summary of parameters*

In their article, the authors of the classification classify printing processes according to the proposed classification parameters summarized in Table 1.1 (Duballet et al. 2017).

An example of classification is the contour crafting process (Khoshnevis et al. 2006): $x_0^{0\text{-}1\text{-}2}$; $x_e^{1\text{-}2}$; e^0; a^1–a^4; s^3; r^0.

– Working scale: the process is designed for printing small buildings. It is also quite precise and can be used to produce small elements. It is therefore x_0^0, x_0^1 or x_0^2, depending on the size of the object to be printed.

– Elemental layer thickness: printed layers are between 1 and 10 cm thick. Depending on the size chosen, the classification is x_e^1, x_e^2.

– Environment: contour crafting is designed for on-site printing. It is therefore e^0.

– Assembly: when building a single-family home, printed elements are used to form larger structures, as well as incorporating lintels and horizontal load-bearing elements (beams) prefabricated elsewhere. Assembly steps a^1 and a^4 are therefore used in this case.

– Use of supports: for single-family home construction, the printer uses prefabricated horizontal supports as supports. These supports play a structural role and are therefore fully integrated into the construction. The support condition is therefore s^3. Some parts of the print do not require support. In this case, condition s^0 is met.

– Robotic complexity: contour crafting uses a crane system that can be likened to a three-axis robot. In this case, robotic complexity is denoted as r^0.

Scale of printed object	x_0^0: <1 m	x_0^1: 1–4 m	x_0^2: 5–10 m	x_0^3: building
Printed material cross-section scale	x_e^0 < 8 mm	x_e^1: 8 mm–5 cm	x_e^2: 5–30cm	x_e^3: >30 cm
Printing environment	e^0: direct on site	e^1: T° and HR control and 0 wind	e^2: prefabrication plant	
Component assembly conditions	a^0: 0 assembly	a^1: Assembling several elements	a^2: Handling elements to place them in their final position	a^3: Assembly of unprinted exterior elements
Use of supports	s^0: No support	s^1: Printed support left in place	s^2: Printed support removed at end of printing	s^3: External support left in place
	s^4: External support removed at end			
Robot complexity	r^0: Three-axis	r^1: Six-axis	r^2: Six-axis rail-mounted	r^3: Six-axis collaborative
	r^4: Six-axis collaborative rail-mounted	r^5: Six-axis rail-mounted	r^6: Six-axis mounted on a three-axis robot	r^7: Six-axis mounted on three-axis robots

Table 1.1. *Classification of printing processes according to (Duballet et al. 2017). Parameters to be taken into account*

1.4.2. *Classification proposed by RILEM*

The work of RILEM's Technical Committee No. 276 has led to the creation of a classification process (Buswell et al. 2020) that encompasses all types of manufacturing processes associated with "digital concrete". This process takes into account a number of principles that allow all automated concrete construction processes to be encompassed.

One of the first principles is considering the printing environment: whether in the prefabrication plant or on-site. Also, the authors drew on pre-existing definitions of material shaping and assembly (Groover 2011) and incorporated standards relating to additive manufacturing (ISO 17296-3 2014; ISO/ASTM 52900 2021). Thus, the classification must account for variations in the size of printed parts and types of cementitious materials used. This classification also describes sub-processes corresponding to the manufacturing technology used (for example, particle beds, extrusion deposition for additive manufacturing). Complementary processes can also be described in the classification (surface treatment, reinforcements, etc.) that may occur sequentially, simultaneously or contiguously.

The proposed classification takes into account:

– the type of material and its service life properties once cured;

– the printing environment (in-plant, on-site, mixed site-plant, on-site but under a protected and regulated area);

– the printed element (size, element to be assembled, resolution and geometric tolerance, etc.);

– process limitations, installation and printing phasing.

The description of a printing process can thus be gathered under the acronym MAPP for Materials, Application, Product, Process, which describes the context and technical details of the printing process.

The processes are classified according to Figure 1.4 reproduced from Buswell et al. (2020). This classification is based on a core group of processes related to automated concrete placement. These include assembly and surface treatment. Sub-processes must be distinct and identified from the main concrete forming and assembly process.

It is noteworthy that a construction process using 3D printing often involves a series of manufacturing and assembly sequences representing the sequencing of each stage, along with elementary processes and sub-processes over time, which enables a clear representation of the studied process and facilitates comparisons with others. Examples of manufacturing processes are shown in Figure 1.5 for a 3D printing process using two-component extrusion/deposition reinforced with continuous glass fibers, known as anisotropic concrete (Bos et al. 2017; Ducoulombier et al. 2020), and another for 3D printing using selective particle bonding (Pierre et al. 2018; Lowke et al. 2018), in line with the classification of processes proposed by RILEM (Buswell et al. 2022).

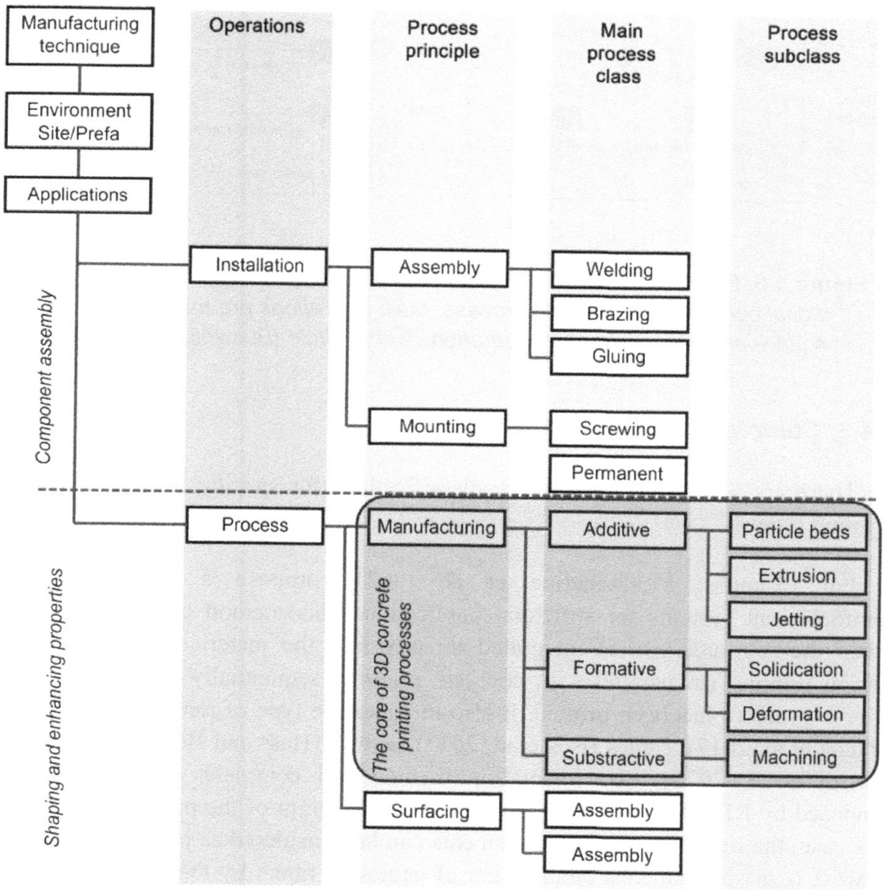

Figure 1.4. *Elementary process classification system proposed by the RILEM DFC technical committee (Buswell et al. 2020, 2022). For a color version of this figure, see www.iste.co.uk/perrot/3dconcrete.zip*

In some cases, a process combines additive manufacturing processes with other formative and subtractive processes involving cast concrete and CNC machine tools for cutting, milling or even welding operations. In such cases, the classification accommodates hybrid processes. This is the case, for example, with the Mesh–Mould process, which is a succession of digital placement of a digital network followed by concrete casting and surface finishing by smoothing, or when a metal reinforcement is welded during concrete 3D printing (Hack and Lauer 2014).

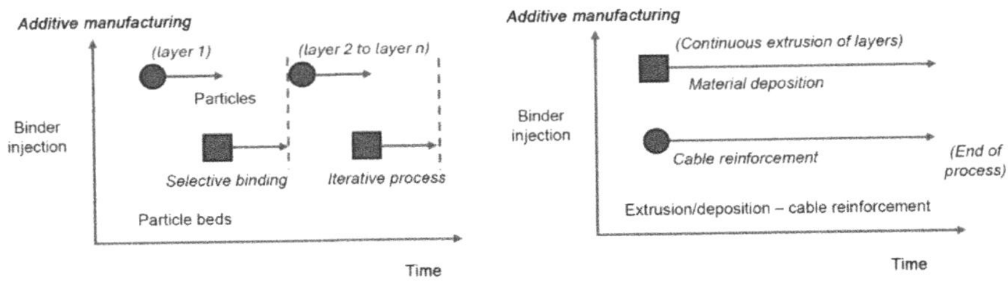

Figure 1.5. *Example of sequencing for a complete description of an automated, digital concrete construction process. Main operations are marked with a square and complementary operations with a circle (Buswell et al. 2020)*

1.4.3. *Complementary classification*

There are several complementary classifications for specific aspects of the 3D printing process.

For example, Mechtcherine et al. (2021) propose a classification of reinforcement systems for structural application. This method takes into account when the reinforcement is integrated directly into the material, before printing, during printing (in parallel with concrete printing, sequentially or alternately) or after a structure has been printed. It also includes the type of reinforcement (fibers (Panda et al. 2017), cables (Bos et al. 2017), screws (Hass and Bos 2020) and nails (Perrot et al. 2020)), the integration method and completes the classification proposed by RILEM concerning the sequencing diagram of the printing process. In this case, the integration of reinforcements can be considered as part of the printing process (e.g. simultaneous entrainment of cables or fibers by the material) or as a sub-process taking place during printing.

Another classification focuses on the print head according to the type of applications and complementary functionalities. This work proposed by Chen et al. (2023) is inspired by the formalism adopted by Duballet et al. (2017).

Finally, classifications of robotic systems and materials have recently been proposed (Sati et al. 2021; Ranjan et al. 2022).

1.5. Printing concrete with alternative binders or without cement?

Concrete printing raises the issue of the carbon impact of printed materials, even if the use of printing means that the structure can be optimized. This carbon impact is all the greater when the quantity of ordinary Portland cement in the composition is high, due to the decarbonation of limestone during cement manufacture and the potential fossil energy used to reach the 1,400°C required to bake the raw materials (Habert et al. 2020).

To ensure material pumpability and adapt to the size of most of the robots used as printers, mortar rather than concrete is used as the printing ink. This increases the quantity of paste and, therefore, of cementitious binder, which carry a significant carbon weight.

Fast-setting cements such as natural prompt cement, calcium aluminate cements or sulfo-aluminate cements are potentially interesting binders to replace traditional cements, as they have a lower carbon weight and high stiffening speeds in line with rapid impression. They are also commonly used as set gas pedals in ordinary Portland cement-based systems. However, researchers have less experience of the durability of these less alkaline systems than traditional cements (Kim et al. 2021; Kleib et al. 2021). It is also important to note that the price of these binders remains higher, which limits their industrial-scale use to date.

The recent NF EN 197-5 standard (Elkhaldi et al. 2023) governing the use of a performance-based approach for low-carbon cements paves the way for the development and use of low-carbon cements containing reduced levels of traditional Portland cement, such as LC3 (limestone–calcined clay–cement). Work is already underway on the use of these new low-carbon binders for 3D printing applications.

The activation of aluminosilicate pozzolanic materials such as blast-furnace slag or calcined clays by more or less concentrated alkaline solutions enables the formation of geopolymers which, in their hardened state, offer strengths of the same order as traditional cement concretes (Duxson et al. 2007). The use of these geopolymers in 3D printing has already been the subject of numerous studies using several different printing methods, such as particle bed printing techniques or the extrusion and filament deposition method (Zhong and Zhang 2022). In these studies, the authors highlight the reactivity of these systems, which enable rapid printing, and the reduced carbon impact associated with the non-use of Portland cement.

For example, a recent study (Zhong and Zhang 2022) indicates a reduction in carbon footprint by a factor of between 2 and 5, and a 50% reduction in the energy required for the printing process compared with printing using cementitious material.

Finally, recent work has focused on the use of soil to create 3D-printed structures as a substitute for cementitious materials in non-structural applications, or where expected levels of mechanical resistance remain low (small buildings, small dwellings, etc.) (Gomaa et al. 2022). 3D soil printing is at a crossroads between several inspirations: robotized methods adapting vernacular techniques such as bauge and pisé, on the one hand (Schweiker et al. 2021), and deposition extrusion printing methods inspired by developments made for cementitious materials, on the other hand (Perrot et al. 2018). It is also important to note that soil can present a wide range of consistencies – fluid, plastic or firm (in this case, the state of the soil is in a so-called wet state as the particles are wetted to a water content insufficient to saturate all the pores between the grains and form a homogeneous paste). For each type of consistency, several manufacturing processes can be imagined, and the techniques developed are manifold, as shown in Figure 1.6.

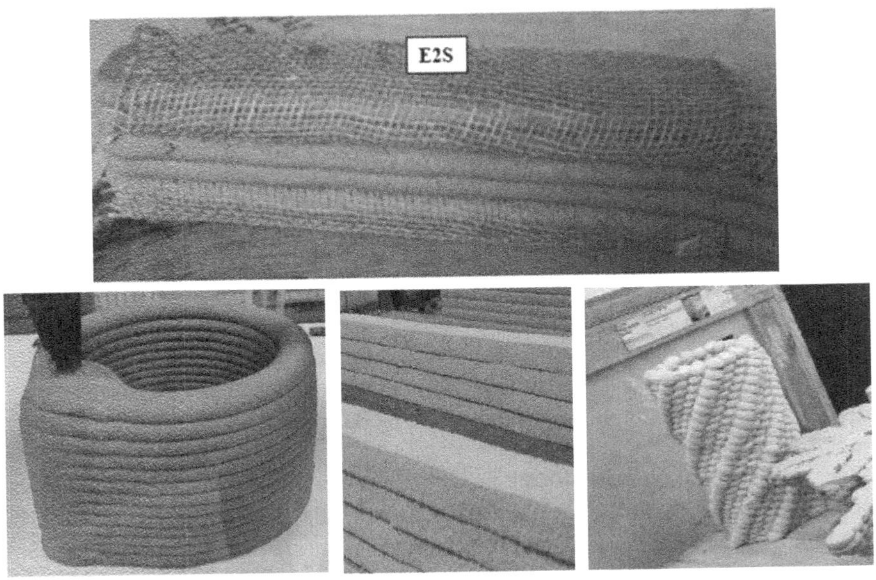

Figure 1.6. *Example of 3D soil printing of structures. For a color version of this figure, see www.iste.co.uk/perrot/3dconcrete.zip*

There are major differences between these two philosophies: the technique inspired by vernacular techniques, such as the robotized pisé technique developed at Braunschweig Technical University, uses only raw soil and no other binders or stabilizers. For this technique, the soil is compacted in a wet state by a robot using a pisoir and mobile formwork. The compaction is sufficient to confer adequate strength to the printed structure, which remains stable and gains in strength as it gradually dries over time, before being put into service.

Techniques inspired by cementitious materials for extrusion printing and filament deposition must use a binder that causes rapid rigidification of the material in order to print quickly (Perrot et al. 2018). This binder can have several origins, cementitious or hydraulic binders (Faleschini et al. 2023), biopolymers such as alginate or hemoglobin extracts (Biggerstaff et al. 2021a, 2021b; Autem et al. 2023). It is also possible to use a binder with thermo-dependent behavior that liquefies on heating for the transport and extrusion phases, then causes the material to stiffen after deposition when it returns to ambient temperature (Christ et al. n.d.; Jacquet and Perrot 2023). In this case, natural waxes or animal gelatins can be used.

1.6. Conclusion

This chapter illustrates the diversity of methods, applications, scales and robotic systems in the field of 3D concrete printing. Today, prefabrication or on-site printing and the manufacture of small parts or entire buildings are all possible realizations by printing.

This multifaceted nature has necessitated a number of classifications, which are also presented in this chapter.

Finally, this chapter explores the possibility of printing alternative materials, which could reduce the carbon impact of printed structures used in construction.

1.7. References

Ahmed, G.H. (2023). A review of “3D concrete printing”: Materials and process characterization, economic considerations and environmental sustainability. *Journal of Building Engineering*, 105863.

ASTM International (2012). ASTM F2792-12a. Standard terminology for additive manufacturing technologies. Standard, ASTM International, West Conshohocken.

Autem, Y., Bourgougnon, N., Guihéneuf, S., Perrot, A. (2023). Comparative study of effects of various seaweed parietal polysaccharides on rheological, mechanical and water-durability properties of earth-based materials. *Materials and Structures*, 56(6), 108.

Barnett, E. and Gosselin, C. (2015). Large-scale 3D printing with a cable-suspended robot. *Additive Manufacturing*, 7, 27–44.

Biggerstaff, A.O., Fuller, G., Lepech, M., Loftus, D. (2021a). Determining the yield stress of a biopolymer-bound soil composite for extrusion-based 3D printing applications. *Construction and Building Materials*, 305, 124730.

Biggerstaff, A.O., Lepech, M.D., Loftus, D.J. (2021b). Evaluation of a biopolymer-bound soil composite for 3D printing on the lunar surface. In *Earth and Space 2021*. ASCE, Reston.

Bos, F.P., Ahmed, Z.Y., Jutinov, E.R., Salet, T.A. (2017). Experimental exploration of metal cable as reinforcement in 3D printed concrete. *Materials*, 10(11), 1314.

Boscaro, F., Quadranti, E., Wangler, T., Mantellato, S., Reiter, L., Flatt, R.J. (2022). Eco-friendly, set-on-demand digital concrete. *3D Printing and Additive Manufacturing*, 9(1), 3–11.

Burger, J., Lloret-Fritschi, E., Scotto, F., Demoulin, T., Gebhard, L., Mata-Falcón, J., Gramazio, F., Kohler, M., Flatt, R.J. (2020). Eggshell: Ultra-thin three-dimensional printed formwork for concrete structures. *3D Printing and Additive Manufacturing*, 7(2), 48–59.

Buswell, R.A., De Silva, W.L., Jones, S.Z., Dirrenberger, J. (2018). 3D printing using concrete extrusion: A roadmap for research. *Cement and Concrete Research*, 112, 37–49.

Buswell, R.A., Da Silva, W.L., Bos, F.P., Schipper, H.R., Lowke, D., Hack, N., Kloft, H., Mechtcherine, V., Wangler, T., Roussel, N. (2020). A process classification framework for defining and describing digital fabrication with concrete. *Cement and Concrete Research*, 134, 106068.

Buswell, R.A., Bos, F.P., da Silva, W.R.L., Hack, N., Kloft, H., Lowke, D., Freund, N., Fromm, A., Dini, E., Wangler, T. (2022). Digital fabrication with cement-based materials: Process classification and case studies. State-of-the-Art Report, RILEM TC 276-DFC.

Carneau, P., Mesnil, R., Roussel, N., Baverel, O. (2020). Additive manufacturing of cantilever – From masonry to concrete 3D printing. *Automation in Construction*, 116, 103184.

Chen, H., Zhang, D., Chen, P., Li, N., Perrot, A. (2023). A review of the extruder system design for large-scale extrusion-based 3D concrete printing. *Materials*, 16(7), 2661.

Christ, J., Perrot, A., Ottosen, L.M., Koss, H. (n.d.). Rheological characterization of temperature-sensitive biopolymer-bound 3d printing concrete. *Construction and Building Materials*, 411, 134337.

Colla, V., Dini, E., Canessa, E., Fonda, C., Zennaro, M. (2013). Large scale 3D printing: From deep sea to the moon. In *Low-Cost 3D Printing, for Science, Education & Sustainable Development*, Canessa, E., Fonda, C., Zennaro, M. (eds). ICTP, Trieste, 127–132.

Construction 3D (n.d.). [Online]. Available at: https://www.constructions-3d.com/.

Craipeau, T., Toussaint, F., Perrot, A., Lecompte, T. (2021). Experimental approach on a moving formwork. *Construction and Building Materials*, 270, 121472.

De Schutter, G., Lesage, K., Mechtcherine, V., Nerella, V.N., Habert, G., Agusti-Juan, I. (2018). Vision of 3D printing with concrete – Technical, economic and environmental potentials. *Cement and Concrete Research*, 112, 25–36.

Dörfler, K., Dielemans, G., Lachmayer, L., Recker, T., Raatz, A., Lowke, D., Gerke, M. (2022). Additive manufacturing using mobile robots: Opportunities and challenges for building construction. *Cement and Concrete Research*, 158, 106772.

Duballet, R., Baverel, O., Dirrenberger, J. (2017). Classification of building systems for concrete 3D printing. *Automation in Construction*, 83, 247–258.

Ducoulombier, N., Demont, L., Chateau, C., Bornert, M., Caron, J.-F. (2020). Additive manufacturing of anisotropic concrete: A flow-based pultrusion of continuous fibers in a cementitious matrix. *Procedia Manufacturing*, 47, 1070–1077.

Duxson, P., Fernández-Jiménez, A., Provis, J.L., Lukey, G.C., Palomo, A., van Deventer, J.S. (2007). Geopolymer technology: The current state of the art. *Journal of Materials Science*, 42, 2917–2933.

Elkhaldi, I., Rozière, E., Loukili, A. (2023). To what extent does decreasing the proportion of clinker in cement production effectively decrease its carbon footprint? *Materials Today: Proceedings* [Online]. Available at: https://www.sciencedirect.com/science/article/pii/S2214785323042190.

Faleschini, F., Trento, D., Masoomi, M., Pellegrino, C., Zanini, M.A. (2023). Sustainable mixes for 3D printing of earth-based constructions. *Construction and Building Materials*, 398, 132496.

Furet, B., Poullain, P., Garnier, S. (2019). 3D printing for construction based on a complex wall of polymer-foam and concrete. *Additive Manufacturing*, 28, 58–64.

Gaël, G., Phelippe, L., Amziane, S. (2023). 3D concrete printer. Google Patents, FR1751938A.

Girmscheid, G. and Moser, S. (2001). Fully automated shotcrete robot for rock support. *Computer-Aided Civil and Infrastructure Engineering*, 16(3), 200–215.

Gomaa, M., Jabi, W., Soebarto, V., Xie, Y.M. (2022). Digital manufacturing for earth construction: A critical review. *Journal of Cleaner Production*, 338, 130630.

Groover, M.P. (2011). *Introduction to Manufacturing Processes*. Wiley, Hoboken.

Habert, G., Miller, S.A., John, V.M., Provis, J.L., Favier, A., Horvath, A., Scrivener, K.L. (2020). Environmental impacts and decarbonization strategies in the cement and concrete industries. *Nature Reviews Earth & Environment*, 1(11), 559–573.

Hack, N. and Kloft, H. (2020). Shotcrete 3D printing technology for the fabrication of slender fully reinforced freeform concrete elements with high surface quality: A real-scale demonstrator. In *Second RILEM International Conference on Concrete and Digital Fabrication: Digital Concrete 2020*. Springer, Heidelberg.

Hack, N. and Lauer, W.V. (2014). Mesh-mould: Robotically fabricated spatial meshes as reinforced concrete formwork. *Architectural Design*, 84(3), 44–53.

Hass, L. and Bos, F. (2020). Bending and pull-out tests on a novel screw type reinforcement for extrusion-based 3D printed concrete. In *Second RILEM International Conference on Concrete and Digital Fabrication: Digital Concrete 2020*. Springer, Heidelberg.

Heidarnezhad, F. and Zhang, Q. (2022). Shotcrete based 3D concrete printing: State of art, challenges, and opportunities. *Construction and Building Materials*, 323, 126545.

ISO (2014). Additive Manufacturing – General Principles. Part 3: Main Characteristics and Corresponding Test Methods. Standard, ISO 17296-3.

ISO (2021). Additive Manufacturing – General Principles – Fundamentals and Vocabulary. Standard, ISO/ASTM 52900.

ISO (2023). Additive Manufacturing for Construction: Qualification Principles – Structural and Infrastructure Elements. Standard, ISO/ASTM 52939.

Izard, J.-B., Dubor, A., Hervé, P.-E., Cabay, E., Culla, D., Rodriguez, M., Barrado, M. (2017). Large-scale 3D printing with cable-driven parallel robots. *Construction Robotics*, 1, 69–76.

Jacquet, Y. and Perrot, A. (2023). Evolutionary approach based on thermoplastic bio-based building material for 3D printing applications: An insight into a mix of clay and wax. In *Bio-Based Building Materials*, Amziane, S., Merta, I., Page, J. (eds). Springer, Cham.

Ji, G., Xiao, J., Zhi, P., Wu, Y.-C., Han, N. (2022). Effects of extrusion parameters on properties of 3D printing concrete with coarse aggregates. *Construction and Building Materials*, 325, 126740.

Keita, E., Bessaies-Bey, H., Zuo, W., Belin, P., Roussel, N. (2019). Weak bond strength between successive layers in extrusion-based additive manufacturing: Measurement and physical origin. *Cement and Concrete Research*, 123, 105787.

Khoshnevis, B. (2004). Automated construction by contour crafting – Related robotics and information technologies. *Automation in Construction*, 13(1), 5–19.

Khoshnevis, B., Hwang, D., Yao, K.-T., Yeh, Z. (2006). Mega-scale fabrication by contour crafting. *International Journal of Industrial and Systems Engineering*, 1(3), 301–320.

Kim, H., Son, H.M., Lee, H.-K. (2021). Review on recent advances in securing the long-term durability of calcium aluminate cement (CAC)-based systems. *Functional Composites and Structures*, 3(3), 035002.

Kleib, J., Aouad, G., Benzerzour, M., Zakhour, M., Abriak, N.-E. (2021). Effect of calcium sulfoaluminate cements composition on their durability. *Construction and Building Materials*, 307, 124952.

Kloft, H., Empelmann, M., Hack, N., Herrmann, E., Lowke, D. (2020). Reinforcement strategies for 3D-concrete-printing. *Civil Engineering Design*, 2(4), 131–139.

Le, T.T., Austin, S.A., Lim, S., Buswell, R.A., Gibb, A.G., Thorpe, T. (2012). Mix design and fresh properties for high-performance printing concrete. *Materials and Structures*, 45, 1221–1232.

Lim, S., Buswell, R.A., Le, T.T., Austin, S.A., Gibb, A.G., Thorpe, T. (2012). Developments in construction-scale additive manufacturing processes. *Automation in Construction*, 21, 262–268.

Lindemann, H., Gerbers, R., Ibrahim, S., Dietrich, F., Herrmann, E., Dröder, K., Raatz, A., Kloft, H. (2019). Development of a shotcrete 3D-printing (SC3DP) technology for additive manufacturing of reinforced freeform concrete structures. In *First RILEM International Conference on Concrete and Digital Fabrication–Digital Concrete 2018*. Springer, Heidelberg.

Lloret-Fritschi, E., Reiter, L., Wangler, T., Gramazio, F., Kohler, M., Flatt, R.J. (2017). Smart dynamic casting: Slipforming with flexible formwork-inline measurement and control. In *HPC/CIC Tromsø 2017*, 27.

Lloret-Fritschi, E., Wangler, T., Gebhard, L., Mata-Falcón, J., Mantellato, S., Scotto, F., Burger, J., Szabo, A., Ruffray, N., Reiter, L. (2020). From smart dynamic casting to a growing family of digital casting systems. *Cement and Concrete Research*, 134, 106071.

Lloret-Fritschi, E., Choma, J., Scotto, F., Szabo, A., Gramazio, F., Kohler, M., Flatt, R.J. (2022). In-crease: Less concrete more paper. *RILEM Technical Letters*, 7, 199–208.

Lowke, D., Dini, E., Perrot, A., Weger, D., Gehlen, C., Dillenburger, B. (2018). Particle-bed 3D printing in concrete construction – Possibilities and challenges. *Cement and Concrete Research*, 112, 50–65.

Lowke, D., Talke, D., Dressler, I., Weger, D., Gehlen, C., Ostertag, C., Rael, R. (2020). Particle bed 3D printing by selective cement activation – Applications, material and process technology. *Cement and Concrete Research*, 134, 106077.

Lowke, D., Vandenberg, A., Pierre, A., Thomas, A., Kloft, H., Hack, N. (2021). Injection 3D concrete printing in a carrier liquid – Underlying physics and applications to lightweight space frame structures. *Cement and Concrete Composites*, 124, 104169.

Lowke, D., Mai, I., Keita, E., Perrot, A., Weger, D., Gehlen, C., Herding, F., Zuo, W., Roussel, N. (2022). Material-process interactions in particle bed 3D printing and the underlying physics. *Cement and Concrete Research*, 156, 106748.

Lowke, D., Anton, A., Buswell, R., Jenny, S.E., Flatt, R.J., Fritschi, E.L., Hack, N., Mai, I., Popescu, M., Kloft, H. (2024). Digital fabrication with concrete beyond horizontal planar layers. *Cement and Concrete Research*, 186, 107663.

Ly, O., Yoris-Nobile, A.I., Sebaibi, N., Blanco-Fernandez, E., Boutouil, M., Castro-Fresno, D., Hall, A.E., Herbert, R.J., Deboucha, W., Reis, B. (2021). Optimisation of 3D printed concrete for artificial reefs: Biofouling and mechanical analysis. *Construction and Building Materials*, 272, 121649.

Mai, I., Lowke, D., Perrot, A. (2022). Fluid intrusion in powder beds for selective cement activation – An experimental and analytical study. *Cement and Concrete Research*, 156, 106771.

Mechtcherine, V., Buswell, R., Kloft, H., Bos, F.P., Hack, N., Wolfs, R., Sanjayan, J., Nematollahi, B., Ivaniuk, E., Neef, T. (2021). Integrating reinforcement in digital fabrication with concrete: A review and classification framework. *Cement and Concrete Composites*, 119, 103964.

Menna, C., Mata-Falcón, J., Bos, F.P., Vantyghem, G., Ferrara, L., Asprone, D., Salet, T., Kaufmann, W. (2020). Opportunities and challenges for structural engineering of digitally fabricated concrete. *Cement and Concrete Research*, 133, 106079.

Moretti, M. (2023). WASP in the edge of 3D printing. In *3D Printing for Construction with Alternative Materials*, Bárbara, R., Ana, S.G., Jorge, L., Leonardo, S. (eds). Springer, Cham.

Nerella, V.N. and Mechtcherine, V. (2019). Studying the printability of fresh concrete for formwork-free concrete onsite 3D printing technology (CONPrint3D). In *3D Concrete Printing Technology*, Sanjayan, J.G., Nazari, A., Nematollahi, B. (eds). Butterworth-Heinemann, Oxford [Online]. Available at: https://www.sciencedirect.com/science/article/pii/B9780128154816000166.

Nicolas, R., Richard, B., Nicolas, D., Irina, I., Temitope, K.J., Dirk, L., Viktor, M., Romain, M., Arnaud, P., Ursula, P. (2022). Assessing the fresh properties of printable cement-based materials: High potential tests for quality control. *Cement and Concrete Research*, 158, 106836.

Panda, B., Paul, S.C., Tan, M.J. (2017). Anisotropic mechanical performance of 3D printed fiber reinforced sustainable construction material. *Materials Letters*, 209, 146–149.

Perrot, A. (2022). M&S highlight: Le et al. (2012), mix design and fresh properties for high-performance printing concrete. *Materials and Structures*, 55(2), 42.

Perrot, A. and Amziane, S. (2019). 3D printing in concrete: General considerations and technologies. In *3D Printing of Concrete: State of the Art and Challenges of the Digital Construction Revolution*, Perrot, A. (ed.). ISTE Ltd, London, and John Wiley & Sons, New York, 1–40.

Perrot, A., Rangeard, D., Pierre, A. (2016). Structural built-up of cement-based materials used for 3D-printing extrusion techniques. *Materials and Structures*, 49, 1213–1220.

Perrot, A., Rangeard, D., Courteille, E. (2018). 3D printing of earth-based materials: Processing aspects. *Construction and Building Materials*, 172, 670–676.

Perrot, A., Jacquet, Y., Rangeard, D., Courteille, E., Sonebi, M. (2020). Nailing of layers: A promising way to reinforce concrete 3D printing structures. *Materials*, 13(7), 1518.

Pierre, A., Weger, D., Perrot, A., Lowke, D. (2018). Penetration of cement pastes into sand packings during 3D printing: Analytical and experimental study. *Materials and Structures*, 51, 1–12.

Pott, A., Meyer, C., Verl, A. (2010). Large-scale assembly of solar power plants with parallel cable robots. In *ISR 2010 (41st International Symposium on Robotics) and ROBOTIK 2010 (6th German Conference on Robotics)*. VDE, Berlin.

Ranjan, R., Kumar, D., Kundu, M., Moi, S.C. (2022). A critical review on classification of materials used in 3D printing process. *Materials Today: Proceedings*, 61, 43–49.

Reiter, L., Wangler, T., Roussel, N., Flatt, R.J. (2018). The role of early age structural build-up in digital fabrication with concrete. *Cement and Concrete Research*, 112, 86–95.

Roussel, N. (2018). Rheological requirements for printable concretes. *Cement and Concrete Research*, 112, 76–85.

Roussel, N. and Coussot, P. (2005). "Fifty-cent rheometer" for yield stress measurements: From slump to spreading flow. *J. Rheol. (1978-Present)*, 49. doi: 10.1122/1.1879041.

Salet, T.A., Ahmed, Z.Y., Bos, F.P., Laagland, H.L. (2018). Design of a 3D printed concrete bridge by testing. *Virtual and Physical Prototyping*, 13(3), 222–236.

Sati, A.S.E., Mantha, B.R., Dabous, S.A., de Soto, B.G. (2021). Classification of robotic 3D printers in the AEC industry. In *ISARC. Proceedings of the International Symposium on Automation and Robotics in Construction*. IAARC Publications, Lille.

Savić, A., Stević, M., Martinović, S., Vlahović, M., Volkov-Husović, T. (2020). Applying concept of 3D printing concrete in wind tower construction. In *Proceedings of the 8th International Conference on Renewable Electrical Power Sources*. Savez mašinskih i elektrotehničkih inženjera i tehničara Srbije (SMEITS), Beograd.

Schack, R., Krause, M., Näther, M., Nerella, V.N. (2017). CONPrint3D: 3D-concrete-printing as an alternative for masonry. *Bauingenieur*, 92, 355–363.

Schweiker, M., Endres, E., Gosslar, J., Hack, N., Hildebrand, L., Creutz, M., Klinge, A., Kloft, H., Knaack, U., Mehnert, J. (2021). Ten questions concerning the potential of digital production and new technologies for contemporary earthen constructions. *Building and Environment*, 206, 108240.

Szabo, A., Reiter, L., Lloret-Fritschi, E., Gramazio, F., Kohler, M., Flatt, R.J. (2020). Mastering yield stress evolution and formwork friction for smart dynamic casting. *Materials*, 13(9), 2084.

Tao, Y., Rahul, A.V., Lesage, K., Yuan, Y., Van Tittelboom, K., De Schutter, G. (2021). Stiffening control of cement-based materials using accelerators in inline mixing processes: Possibilities and challenges. *Cement and Concrete Composites*, 119, 103972.

Tao, Y., Rahul, A.V., Lesage, K., Van Tittelboom, K., Yuan, Y., De Schutter, G. (2022). Mechanical and microstructural properties of 3D printable concrete in the context of the twin-pipe pumping strategy. *Cement and Concrete Composites*, 125, 104324.

Tay, Y.W.D., Li, M.Y., Tan, M.J. (2019). Effect of printing parameters in 3D concrete printing: Printing region and support structures. *Journal of Materials Processing Technology*, 271, 261–270.

Vantyghem, G., De Corte, W., Shakour, E., Amir, O. (2020). 3D printing of a post-tensioned concrete girder designed by topology optimization. *Automation in Construction*, 112, 103084.

Volpe, S., Sangiorgio, V., Petrella, A., Coppola, A., Notarnicola, M., Fiorito, F. (2021). Building envelope prefabricated with 3D printing technology. *Sustainability*, 13(16), 8923.

Wangler, T., Lloret, E., Reiter, L., Hack, N., Gramazio, F., Kohler, M., Bernhard, M., Dillenburger, B., Buchli, J., Roussel, N. (2016). Digital concrete: opportunities and challenges. *RILEM Technical Letters*, 1, 67–75.

Wangler, T., Pileggi, R., Gürel, S., Flatt, R.J. (2022). A chemical process engineering look at digital concrete processes: Critical step design, inline mixing, and scaleup. *Cement and Concrete Research*, 155, 106782.

Weger, D., Pierre, A., Perrot, A., Kränkel, T., Lowke, D., Gehlen, C. (2021). Penetration of cement pastes into particle-beds: A comparison of penetration models. *Materials*, 14(2), 389.

Xiao, J., Ji, G., Zhang, Y., Ma, G., Mechtcherine, V., Pan, J., Wang, L., Ding, T., Duan, Z., Du, S. (2021). Large-scale 3D printing concrete technology: Current status and future opportunities. *Cement and Concrete Composites*, 122, 104115.

Zhang, X., Li, M., Lim, J.H., Weng, Y., Tay, Y.W.D., Pham, H., Pham, Q.-C. (2018). Large-scale 3D printing by a team of mobile robots. *Automation in Construction*, 95, 98–106.

Zhang, K., Chermprayong, P., Xiao, F., Tzoumanikas, D., Dams, B., Kay, S., Kocer, B.B., Burns, A., Orr, L., Choi, C. (2022). Aerial additive manufacturing with multiple autonomous robots. *Nature*, 609(7928), 709–717.

Zhong, H. and Zhang, M. (2022). 3D printing geopolymers: A review. *Cement and Concrete Composites*, 128, 104455.

Zuo, W., Dong, C., Belin, P., Roussel, N., Keita, E. (2022). Capillary imbibition depth in particle-bed 3D printing – Physical frame and one-dimensional experiments. *Cement and Concrete Research*, 156, 106740.

2

3D Concrete Printing by Extrusion and Filament Deposition

Arnaud PERROT[1] and Yohan JACQUET[2]

[1] *IRDL, Université de Bretagne Sud, Lorient, France*

[2] *Technische Universität Berlin, Germany*

2.1. Introduction

Extrusion/deposition 3D printing is currently the most widely studied and most widespread process in concrete additive manufacturing (Wangler et al. 2016; Mechtcherine et al. 2020).

This robotized manufacturing process can be broken down into a number of stages, from the manufacture of the material to the curing of the printed object to ensure good mechanical performance. This complex process includes conveying the material to the print head, often by pumping, shaping it by extruding a filament material, then depositing it in successive layers to form a stable structure (Perrot et al. 2016, p. 201; Roussel 2018; Wolfs et al. 2018).

For each of these steps, the fresh cementitious materials must exhibit adequate rheological behavior to enable them to move on to the next step. In particular, the material must be able to flow to the printing nozzle and must be able to maintain a shape after deposition to ensure the stability of the structure during printing.

Generally speaking, in the laboratory or on site, a print head attached to a robotized system deposits the fed material, layer by layer, following a programmed trajectory to form the structure to be printed (Figure 2.1).

Depending on the application and the form to be printed, there is a wide variety of extrusion-deposition printing methods. These differ in terms of delivery method, type of formulation (single-component or two-component with the addition of a structuring/flocculating agent or hydration accelerator at the printhead) (Mechtcherine et al. 2020; Perrot et al. 2021; Wangler et al. 2022) and the robotic system used. For the last category, various technical solutions can move the print head, such as gantries or overhead cranes (Khoshnevis 2004), fixed articulated arms (Gaudillière et al. 2020) or mounted on rails or mobile machinery (Khoshnevis et al. 2005; Leach et al. 2012), delta robots (Asprone et al. 2018), or supports or even traditional lifting means diverted from their original purpose (Schack et al. 2017; Nerella and Mechtcherine 2019).

This diversity of robotic solutions and printable materials makes it possible to build on a variety of scales from small prefabricated elements to entire buildings (Souza et al. 2020; Xiao et al. 2021).

To understand the 3D concrete printing process by extrusion-deposition, we must be familiar with the behavior of the material in its fresh state. This will enable us to understand each stage of the printing process and the constraints imposed on the material by the printing process.

Figure 2.1. *3D printing of mortar by extrusion and filament deposition using a pump and a six-axis robotic arm. For a color version of this figure, see www.iste.co.uk/perrot/3dconcrete.zip*

In this chapter, the diversity of extrusion-deposition printing processes will be presented to provide a comprehensive overview of the range of possibilities. We will then focus on cementitious materials in their fresh state to understand their behavior during each stage of the printing process, from mixing the materials to the evolution of the printed structure during the first hours and days after printing.

2.2. Major printing families

2.2.1. *A matter of scale*

The size of the printed structure has a direct impact not only on the means of printing, but also on the properties of the material to be printed.

Printing small parts, for example, of the order of a few tens of centimeters in height, implies a limited load on the fresh material during printing and therefore does not necessarily require the cementitious material to harden rapidly for the structure to remain stable. In fact, the weight acting on the base layer depends directly on the number of layers stacked and the height of the printed form. This printing situation is the simplest from a material point of view, as it is compatible with common mortars and concretes.

On the other hand, a prefabricated part such as a lost column formwork (around 3-meters high for a standard building level height) will induce a load of around 70 kPa (a hydrostatic-type load equal to the product of density, printed height and acceleration of gravity). What is more, the length of the bead to be printed for one layer is small, and the rate of rise of the structure can be very rapid (of the order of 10 cm or more per minute). Traditional cementitious material will not be able to withstand the induced loading and ensure the stability of the structure. It is therefore necessary to accelerate the curing rate of the cementitious binder to guarantee successful printing. Formulating a material with texturizing agents or setting accelerators is therefore necessary in this case.

A third printing situation is the automated manufacture of entire buildings. In this case, the bead length of a layer can be several tens of meters, and the speed at which the structure rises during printing is fairly low. In this case, the formulation of the material to be printed does not necessarily require high curing speeds to ensure that the structure remains stable during printing. On the contrary, too fast a curing speed could be detrimental to good inter-layer bonding, and therefore to the mechanical strength of the printed structure once cured. It is important to be able to guarantee the robustness of the material concerning exposure to the external environment between two successive deposits. This can be achieved by adding additives to protect the material from over-drying or leaching when exposed to rain.

A comparison of these three situations shows that the application, and particularly the size of the printed element, has major repercussions on the formulation of the printed cementitious material and its characteristics in the fresh state (in particular its setting speed). In order to guarantee the success of the printing process in relation to the intended application, it is therefore possible to influence not only the formulation of the material, but also the robotic solution used.

2.2.2. *Single- and two-component materials*

When it comes to materials used in 3D printing by extrusion and filament deposition, there are two strategies based on their formulation.

The first material solution is known as single component and is based on the formulation of a cementitious material that can be pumped to the print head and has a structural build-up speed adapted to the printed form. This method has the advantage of being relatively simple from a technical point of view, as the cementitious material undergoes no change between the end of the mixing operation and the deposition. However, this method requires extreme precision in terms of print sequencing, as the age of the material is directly related to its strength level. Thus, any pause in the process can induce a delay leading to hardening of the material before it is deposited either in the tank or in the pipes of the concrete pump. In extreme cases, this can lead to pump blockage and printing stoppage. This lack of process robustness becomes more significant the higher the structural build-up speed. Single-component materials are therefore particularly suitable for printing situations where the required structural build-up rate remains limited. However, it is worth noting that some applications requiring rapid structural build-up speeds can be achieved with single-component materials, as shown by Weber beamix or Vertico, for example, but this requires mastery of the material and the process, as well as considerable technical expertise.

The second solution involves mixing two materials at the print head, transforming a pumpable material with stable rheology over time into a rapidly hardening material. This formulation strategy increases the robustness of the process with regard to the pumping stage (little risk of material hardening in the pipe), but requires the choice and dosage of a texturizing agent or setting accelerator to achieve a high structuring speed, and the design of an active or passive system for injecting this agent and mixing to achieve a homogeneous material (Reiter et al. 2020; Wangler et al. 2022). This formulation stage can be tricky because hydration kinetics are highly dependent on the quality and composition of the raw materials. The design of the injection and mixing process can therefore be complex and also requires dimensioning the size of the mixing zone, the active or passive mixing system, and ensuring sufficient residence time to obtain adequate mixing (Wangler et al. 2020).

Traditionally, for this so-called two-component process, the cementitious material is pumped and a small quantity of accelerator is added at the print head. Recently, a dual pumping system where the accelerator mixed with an inert suspension (sand and fine limestone) is pumped in parallel with an equivalent quantity of cementitious material before the materials are mixed at the printhead (Tao et al. 2022). This method allows for good control of accelerator dosage and delivers excellent results.

Two-component materials are particularly well-suited to the 3D printing of slender prefabricated parts or parts with cantilevers that need to be strengthened very quickly and are frequently used by prefabricators of printed parts.

2.2.3. *Robotic complexity*

The complexity of the robot and the number of its degrees of freedom will have a direct impact on the extrusion-deposition printing process. The simplest three-degree-of-freedom robots (Cartesian robots) generally feature a cylindrical cross-section printing nozzle, resulting in a surface structure with strands.

The addition of a degree of freedom, rotation around the vertical axis, makes it possible to print with a rectangular-section nozzle, producing a printed structure with a potentially smooth, bulge-free surface. In this case, the robot's trajectory must be programmed so that the printhead cross-section remains perpendicular to the robot's direction of travel.

Finally, the use of a six-axis robot means that complex trajectories can be achieved with a vertically oriented print head, enabling cantilevered structures to be produced. The progressive orientation of the print head as the structure is printed and raised allows the shear strength of the interfaces to be mobilized for local and global stability during printing. This robotic solution, more complex from a trajectory generation point of view, is therefore to be preferred when printing complex structures (Gosselin et al. 2016; Carneau et al. 2020). It should be noted that some authors have drawn inspiration from traditional masonry techniques for the manufacture of vaults and domes, utilizing six-axis robots in a concrete 3D printing context (Carneau et al. 2020).

2.3. Printable materials

2.3.1. *Material behavior during printing*

Cementitious materials are yield stress materials (Tattersall 1976; Tattersall and Banfill 1983). This means that the material only begins to flow when subjected to a stress above a critical value known as the plastic yield stress. This threshold value depends on the type of loading and differs in shear, elongation (induced by a difference in axial stresses) or combined (shear + elongation) flow. Below this critical value, it is often assumed that cementitious materials behave as elastic solids (Roussel 2018; Mechtcherine et al. 2020), although the material behavior has a viscous component (Esposito et al. 2021). It is important to note that it is on this duality of behavior that the possibility of printing a cementitious material rests: it is necessary to impose a stress greater than the yield stress to convey the material to its deposition and then ensure that the stresses acting on the structure during printing remain below the material's flow yield stress (Perrot et al. 2016). The elastic deformations of the loaded material must also be controlled below its shear yield stress to ensure good shape control and avoid buckling of the structure (Wolfs et al. 2018).

To facilitate printing, it is important to note that we can take advantage of the rigidification of the material linked to the chemical activity of the cement to increase the plastic yield stress and rigidity of the deposited layers (Roussel 2006, 2018).

When printing mortar or concrete, the material must, depending on the stage, exhibit either solid behavior to hold its shape (after deposition) or liquid behavior to flow and be conveyed (during pumping and extrusion). It is therefore essential to understand the transition between the liquid and solid behavior of the material. This transition is a decisive characteristic of fresh concrete for printing purposes.

The so-called static yield stress, which is a stress that occurs at a critical deformation called ε_c in elongation or γ_c in shear. This flow or plasticity yield stress is τ_0 for shear flow and σ_0 for elongation flow.

For complex loads, before the cement paste or mortar sets, a Von Mises plasticity criterion can be used (Mettler et al. 2016) to check whether the material will behave as an elastic solid (criterion not met) or as a fluid (criterion met).

This criterion is written in the following form:

$$\sigma_{eq} = \frac{1}{\sqrt{2}}\sqrt{\begin{matrix}(\sigma_{11}-\sigma_{22})^2+(\sigma_{33}-\sigma_{22})^2+(\sigma_{33}-\sigma_{11})^2\\+6(\tau_{12}{}^2+\tau_{13}{}^2+\tau_{23}{}^2)\end{matrix}}$$

$$\sigma_{eq} \leq \sigma_0 \qquad [2.1]$$

Here, σ_{ii} and τ_{ij} are the components of the stress tensor, σ_{eq} is the equivalent Von Mises stress and σ_0 is the value of the plastic yield stress.

Other criteria can be used, such as the Tresca criterion, although the Von Mises criterion is more widely used for cementitious materials in the fresh state (Adams et al. 1997; Ovarlez and Roussel 2006; Perrot et al. 2012).

For unconfined compressive loading along axis 1, as in a vertically printed layer, the Von Mises criterion becomes:

$$\sigma_{eq} = \sigma_{11}, \sigma_{11} \leq \sigma_0 \qquad [2.2]$$

It should be noted that the use of these criteria makes it possible to determine a link between the shear yield stress and the elongation yield stress. Thus, for the Von Mises criterion in simple shear (axes 1 and 2), we can write:

$$\sigma_{eq} = \sqrt{3}\tau_{12}, \tau_{12} \leq \frac{\sigma_0}{\sqrt{3}} = \tau_0 \qquad [2.3]$$

This load is representative of what happens when the material is flowing in a pipe. τ_0 is also known as the (static) shear yield stress. This is the shear yield stress that is traditionally used in many concrete-related manufacturing processes to qualify consistency and suitability for use. This static shear yield stress is representative of the state of structural build-up of the material's microstructure, which develops resistance over time.

For formulations containing a high solid fraction close to the maximum compactness of the granular mix, the material's behavior becomes more complex and pressure-sensitive (Mettler et al. 2016; Wolfs et al. 2019; Jacquet et al. 2021a). In this case, soil mechanics models such as the Coulomb or Drucker–Prager criteria can be used to determine material flow and the onset of plastic deformation (Khelifi et al. 2013). Coulomb's model uses an internal angle of friction to account for the material's resistance to inter-grain friction.

Finally, we must bear in mind the multiphase nature of cementitious materials and the possible phase separations that may occur during the different phases of the process in order to take advantage of them or to avoid possible inconveniences (Khelifi et al. 2013; Perrot and Rangeard 2016; Massoussi et al. 2017).

2.3.2. *Material flow behavior*

Under steady-state flow conditions, cementitious materials flow as a quasi-Newtonian viscous fluid (viscosity η) or as a power-law fluid when subjected to a stress above the so-called dynamic yield stress τ_0 (Roussel 2005; Roussel et al. 2010). This type of behavior can be described by a Bingham model written as equation [2.4] or by the Herschel–Bulkley model written as equation [2.5] for shear stress (as traditionally written in the literature):

$$\tau = \tau_0 + \eta\dot{\gamma}, \qquad [2.4]$$

$$\tau = \tau_0 + \mu\dot{\gamma}^n \qquad [2.5]$$

Here, $\dot{\gamma}$ is the shear rate, τ is the shear stress, μ is the consistency for the Herschel–Bulkley model and n is the flow index. The value of n controls the decrease or increase in apparent viscosity with increasing shear stress. The behavior is said to be shear-thinning for $n < 1$ or shear-thickening if $n > 1$, as illustrated in Figure 2.5.

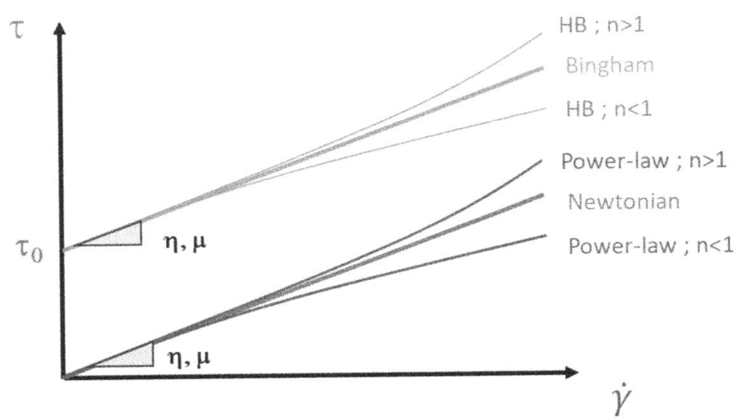

Figure 2.2. *Shear stress–shear rate relationship for a Bingham fluid (dashed linear relationship) and two Herschel–Bulkley (HB) fluids with flow indices n less than and greater than 1. For a color version of this figure, see www.iste.co.uk/perrot/3dconcrete.zip*

Models describing the flow behavior of cementitious materials are useful for simulating the flows involved in pumping and extruding cementitious materials when the flow is steady state and the material's microstructure does not change due to colloidal interactions or the chemical setting of the cement. In this case, the cement grain network is said to be destructured and the flow is said to be steady state.

2.3.3. *Material behavior at rest*

2.3.3.1. *Elastic deformation below the yield stress*

Under yield stress loading, such as the loading acting on a layer of material deposited during printing, it is commonly accepted that cementitious materials behave as linear elastic solids exhibiting an elastic modulus E in compression and G in shear. This modeling is undoubtedly too simplistic, as behavior below the shear yield stress is clearly nonlinear (Roussel et al. 2019). Also, deformations are not fully reversible due to a viscous component in the material's behavior below the shear yield stress (Esposito et al. 2021). However, the use of a linear model describing the material's apparent behavior is sufficient to describe the material's behavior in a printing situation (Wolfs et al. 2018; Suiker et al. 2020).

This elastic behavior is therefore important for calculating the deformations of printed structures under their own weight.

The critical elastic deformation is often used to define the limit of the elastic domain. This deformation is written ε_c in compression/tension or γ_c in shear. The ratio between the yield stresses, in compression and shear, and the corresponding critical deformations is used to define the corresponding apparent elastic stiffness moduli $E = \sigma_0/\varepsilon_c$ in compression and $G = \tau_0/\gamma_c$ in shear (Roussel et al. 2010; Roussel 2018).

2.3.3.2. *Evolution of material structure at rest*

Once in contact with water, cement undergoes a temporal evolution that changes its physical and rheological characteristics, as well as the structure of the suspensions. It is this progressive modification of the microstructure linked to cement reactivity that results in an increase in the material's strength and rigidity, which is mobilized to enable the structure being printed to remain stable. In fact, unlike plastics, the material's behavior cannot be considered thermo-dependent, and the rigidification of the material is not due to cooling. Matching the kinetics of mechanical structural build-up of the material linked to the activity of cement in water to the printing rate is therefore crucial to the success of printing (Wangler et al. 2016; Perrot and Rangeard 2016).

To explain this rigidification, we need to take into account the organization of the network of cement grains suspended in water, as shown in Figure 2.3, which illustrates the different phases in the evolution of the microstructure. After intense mixing inducing destructuring, cement grains begin to flocculate under the influence of colloidal interactions. This flocculation into a structured network stiffens and strengthens the cementitious material on timescales of tens of seconds (Roussel et al. 2012; Roussel 2018; Kruger et al. 2019).

In a second phase, the rapid formation of CSH bridges in the contact zones between the grains is responsible for further mechanical structural build-up of the resting cementitious material. This phenomenon of cement grain nucleation occurs during the first few minutes or even tens of minutes after water-cement contact, before setting during the so-called dormant period (Sleiman et al. 2010). Hydrates formation is chemically irreversible, but bridges between grains can be physically broken if the pumping and/or mixing system is sufficiently powerful. A true loss of workability (irreversible change in behavior) is achieved when the rate of hydrate formation becomes too rapid (Mantellato et al. 2019).

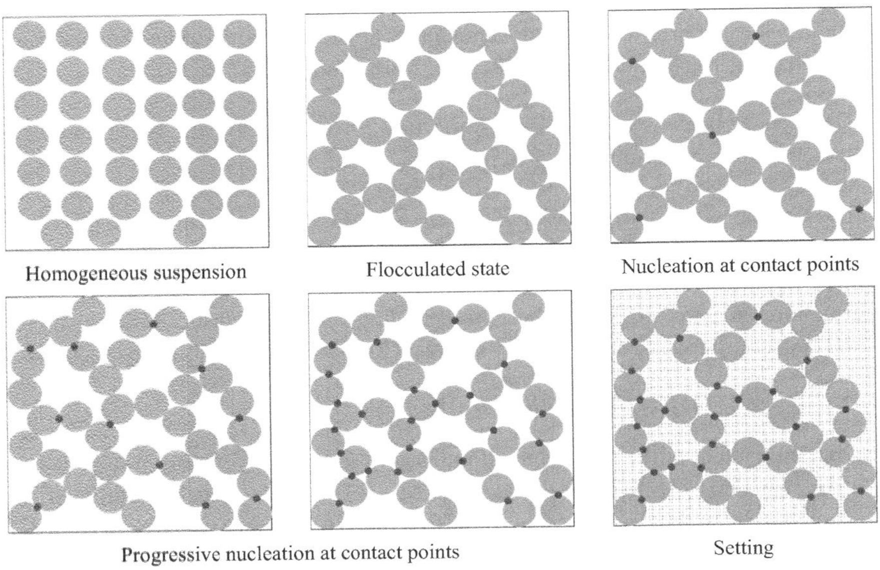

Figure 2.3. *Evolution of the microstructure of a cement suspension at rest. Large circles represent cement grains, small circles hydrates. For a color version of this figure, see www.iste.co.uk/perrot/3dconcrete.zip*

These microstructural changes result in an increase in the material's behavioral parameters with increasing dwell time. To study the various stages of the printing process, researchers often use the evolution of stiffness and shear yield stress. The evolution of the material's shear yield stress at rest is the most widely used indicator to describe the evolution of microstructure on mechanical behavior.

The first way of modeling the increase in the shear yield stress with time is a linear model that defines an A_{thix} structural build-up rate in Pa·s^{-1} over a given period (Roussel 2005, 2006). In this case, the equation is written as follows:

$$\tau_0(t) = \tau_{0,t=0} + A_{thix}t \quad [2.6]$$

Here, $\tau_{0,t=0}$ is the material's shear yield stress in the destructured state and t is the duration of the material's rest period. In this case, it is possible to define a characteristic structural build-up time t_{2x}, which corresponds to the time required to double the material's yield stress. This characteristic time is:

$$t_{2x} = \tau_{0,t=0}/A_{thix} \quad [2.7]$$

Linear modeling of the shear yield stress is generally valid over a given resting period of the cementitious material (Subramaniam and Wang 2010) of a few tens of minutes for conventional non-accelerated cements. Beyond this, the increase in strength accelerates and the structural build-up kinetics become exponential (Perrot et al. 2015; Lecompte and Perrot 2017). This change in rate can be explained by an increase in the rate of formation of hydrates that interact with each other. Perrot et al. (2015) proposed an exponential shear yield stress evolution law that tends toward the linear model for short times. This model is written as:

$$\tau_0(t) = A_{thix}t_c\left(e^{t/t_c} - 1\right) + \tau_{0,t=0} \quad [2.8]$$

Here, t_c is a characteristic time for which the behavior can be considered linear. Figure 2.4 shows that the model in equation [2.8] can be used to describe the evolution of the shear yield stress over longer periods than the linear model.

However, as explained above, the evolution of the shear yield stress is due to at least two phenomena: the flocculation of colloidal particles and the creation of hydrates, which create new interactions and then increase the solid volume fraction in the material. As a result, Kruger et al. (2019) have developed a bilinear model that dissociates these two phenomena by defining two characteristic velocities, a flocculation velocity R_{thix} and a cement grain nucleation velocity A_{thix}. The model is written as:

$$\begin{cases} t < t_{rf}; \tau_0 = \tau_{0,t=0} + R_{thix}.t_{rf} \\ t > t_{rf}; \tau_0 = \tau_{0,t=0} + R_{thix}.t_{rf} + A_{thix}.(t - t_{rf}) \end{cases} \qquad [2.9]$$

The bilinear model provides a good description of the evolution of the shear yield stress in the first instants when the flocculation phenomenon has not ended. As proposed by Nicolas et al. (2022), it is possible to combine the Perrot and Kruger models to obtain a more accurate description of the evolution of the shear yield stress over longer periods.

There are other, more sophisticated models to describe the evolution of the shear yield stress over time (Mettler et al. 2016; Lecompte and Perrot 2017; Wolfs et al. 2019).

The creation of new hydrates during the material's resting period induces a change in the type of behavior of the cementitious material, which becomes increasingly granular and then brittle solid.

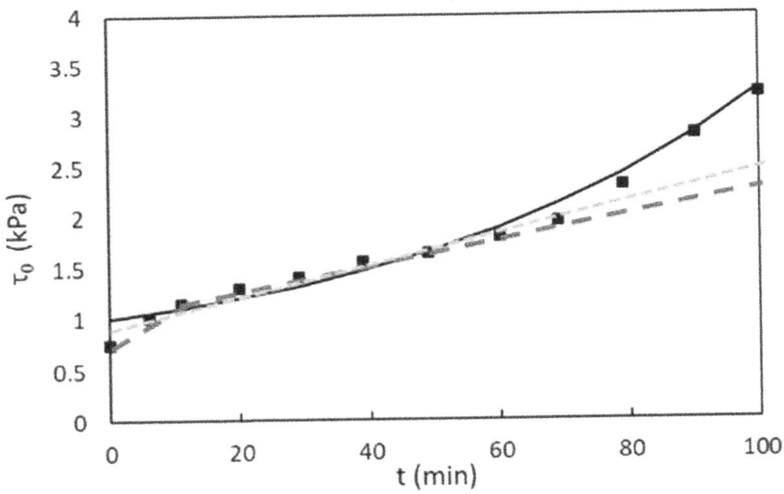

Figure 2.4. *Example of applications of models describing the evolution of the shear yield stress of a mortar at rest. The solid line corresponds to the exponential model, the dashed line to the bilinear model, the dotted line to the linear model and the dots to the experimental measurements. For a color version of this figure, see www.iste.co.uk/perrot/3dconcrete.zip*

Some authors have thus demonstrated that the Von Mises-type plasticity criterion is no longer adequate to describe material failure. Mechanical behavior thus shows a progressive tension/compression asymmetry, leading to brittle behavior typical of concrete with tensile weakness (Jacquet et al. 2021b). In this case, a Coulomb-type plasticity criterion is used to model the pressure-sensitive failure of the material.

In terms of material stiffness and pseudo-elastic deformation, the critical strain decreases slightly as the material stiffens (Mettler et al. 2016; Roussel 2018).

This increase in stiffness over time translates into an increase in elastic modulus. These increases in stiffness and strength enable the deposited material to withstand the increased loads associated with printing the structure. In this way, it is possible to calculate and predict the optimum manufacturing speeds to guarantee the structure's stability and ensure compensation for elastic deformation, as we shall see in the following sections.

2.3.4. *Printing specifications*

The behavior of cementitious materials in the fresh state, with both solid and evolving behavior at rest and fluid behavior above a certain level of stress, makes it possible to consider material printing.

At each stage of the printing process, the material behavior will be specified (Mechtcherine et al. 2020). In the next section, the various stages of the printing process will be presented and described in order to understand the material parameters involved and to move toward a specification of characteristics.

2.4. The main stages in 3D extrusion printing

2.4.1. *Mixing*

When 3D printing elements in cementitious materials, mixing is a crucial stage to ensure that the material has the expected properties, both in its fresh and hardened state. Indeed, mixing energy has a direct influence on the hardened properties of concrete, particularly when numerous admixtures are used, which need to be well dispersed in the material to make the mixture homogeneous (Dils et al. 2012).

Monitoring the power consumed by the mixer, and the appearance of a consumption plateau, is a sign of efficient mixing and that the mixture has become homogeneous (Chopin et al. 2004). However, it should be noted that power consumption data are often very irregular and can be difficult to interpret. The use of tracers or colorimetric analysis can prove effective in ensuring material homogeneity. For example, Jézéquel and Collin (2007) have shown that the colorimetric index of tracer particles tends toward a plateau value following exponential kinetics.

$$c = c_{plateau} + (c_0 - c_{plateau})\, exp\left(\frac{-t}{t_c}\right), \qquad [2.10]$$

where $c_{plateau}$ is the final colorimetric index, c_0 is the initial value and t_c is the characteristic dispersion time. It should be noted that the characteristic time takes into account the various mechanisms involved in mixing. This characteristic time depends on the speed of the mixer and the rheology of the material and varies between 1 and 10 minutes (Jézéquel and Collin 2007).

For 3D printing applications, it has been demonstrated that intense mixing under a high shear gradient can generate cement grain crushing, which creates colloidal nanoparticles that promote the material's mechanical structural build-up speed and binder reactivity (Vandenberg et al. 2019). This method, which requires fine control of the process, has yet to be adapted to industrial scale.

Finally, in two-component printing systems, a mixing system is needed to inject a "hardener" additive for rapid printing (Marchon et al. 2018). This hardener, perhaps a setting accelerator, a fast-setting binder or a colloidal agent, promotes the formation of a solid flocculated network. In such a printhead, the quality of the mixture will depend on the residence time of the material, which can vary from one to several tens of seconds (Wangler et al. 2020). The use of a static or dynamic in-line mixing system is necessary to ensure material homogeneity. This mixer must be able to create sheared layers spaced by a length smaller than the characteristic diffusion length of the material during its residence time in the print head, which is below 100 μm (Mechtcherine et al. 2020; Perrot et al. 2021).

Recent studies present a state of the art of printhead and in-line mixer technologies and highlight the need for prediction and control of the quality of the mix achieved (Chen et al. 2023).

2.4.2. *Pumping*

Pumping is a classic step in the manufacturing process for concrete and other cementitious mixes. Large-scale on-site pumping with printable concretes is not yet a common problem, but most printing processes use a pump to convey the material over distances of around ten meters or so. However, pumpability over medium distances is a prerequisite for many printing projects, even in the factory. The pumpability of concrete depends on materials and processing parameters, and it is generally described by pressure–flow relationships (Kaplan et al. 2005; De Schutter and Feys 2016; Feys et al. 2022).

It is therefore necessary to adapt the power and pumping system to the formulation used. For example, the diameter of the aggregates must be well suited to the operation of the pressurization system and to the diameter of the pipes to prevent blockage. Similarly, the dosage of aggregates should be limited so as not to give the cementitious material a predominantly frictional character that could lead to flow stoppage.

Pipe flow in a viscoplastic material such as concrete is characterized by a shear velocity gradient that decreases with distance from the pipe wall, and an unsheared zone in the center of the pipe. This is because the shear stress in this zone is below the material's τ_0 shear yield stress.

However, during pumping, a phenomenon of particle migration under shear induces a movement of particles toward the less sheared zones (in the center of the pipe) and generates the formation of a lubrication layer composed mainly of cement paste and the finest sand grains (Secrieru et al. 2018; Tavangar et al. 2022). This phenomenon must be taken into account when predicting pumping pressure and flow rate, and it facilitates the flow of cementitious material. The behavior of the material in the pipe is then dictated by these wall effects and the rheological characteristics of this lubrication layer, which has a shear yield stress τ_{0i} and a viscosity μ_i lower than the values of the initial material.

Kaplan's (1999) work has made it possible to construct relationships between pumping flow and pressure for cementitious materials with this lubrication layer. The pressure ΔP and flow rate Q_p are linked by equations using the parameters of the Bingham model of the concrete and lubrication layer, and the geometry of the pumping pipe (length L_{pipe} and radius R_{pipe}). Equation [2.11] is applicable in the case of a flow where only the lubrication layer is sheared. The adjustment coefficient k can be obtained by means of tribological measurements (Kaplan et al. 2005; Kwon et al. 2016).

$$\Delta P = \frac{2L_{pipe}}{R_{pipe}}\left[\frac{Q_p\mu_i}{\pi R_{pipe}^2 k} + \tau_{0,i}\right] \qquad [2.11]$$

Equation [2.12] covers the case of flow where shear occurs in the lubrication layer and within the concrete:

$$\Delta P = \frac{2L_{pipe}}{R_{pipe}}\left[\frac{\frac{Q_p}{\pi R_{pipe}^2 k}-\frac{R_{pipe}}{4\mu}\tau_{0,i}+\frac{R_{pipe}}{3\mu}\tau_0}{1+\frac{R_{pipe}}{4\mu}\mu_i}\mu_i+\tau_{0,i}\right] \quad [2.12]$$

Determining the rheological properties of the concrete and lubricating layer, as well as the thickness of the lubricating layer, is a difficult task that limits the applicability of the above models for predicting pumping pressures (Choi et al. 2014). In real cases, the use of numerical simulation tools may be necessary to properly account for particular geometries such as bends, narrowings or widenings of sections (De Schryver et al. 2021).

2.4.3. *Extrusion*

Extrusion is the last phase in the pre-deposition routing of the cementitious material filament. Passage through the printhead die corresponds to the extrusion phase and gives the bead its shape before deposition (Perrot et al. 2019). Extrusion will therefore induce a narrowing of the material filament cross-section, necessitating an increase in the pumping pressure required to ensure material flow in the printing system.

In the filament deposition/extrusion printing process, extrusion occurs as a continuation of pumping. As in the pump lines, the advancement of the material through the extruder is often influenced by the formation of a lubrication boundary layer where shear is concentrated (Khelifi et al. 2013). This condition generates a so-called plug flow, in which the material is only sheared very locally at the interface.

Extrusion flow can be broken down by a flow description into three distinct parts (Perrot et al. 2012) (Figure 2.5):

– A plug flow localized in the extruder body where the paste is sheared off the extruder wall. This slippage induces a shearing force that must be overcome to ensure flow.

– A conical "dead" zone where the material remains static around the die exit. It acts as a progressive conical constriction and can be eliminated if the extruder die itself has a progressive constriction of optimized shape.

– A shaping zone located in the tapered narrowing region of the die section.

This zonal decomposition allows us to write the extrusion force as the sum of two contributions relative to the different parts of the flow:

– Several models have been developed to predict shaping force as a function of die geometry and the rheological parameters of the cementitious material (Benbow and Bridgwater 1993; Perrot et al. 2012; Zhou et al. 2013). This effort is denoted as F_{pl}.

– Shear stress at the wall is often assimilated to parietal friction F_{fr}. This force is linked to the parietal stress τ_w, which depends on the lubricating layer's behavior at the interface.

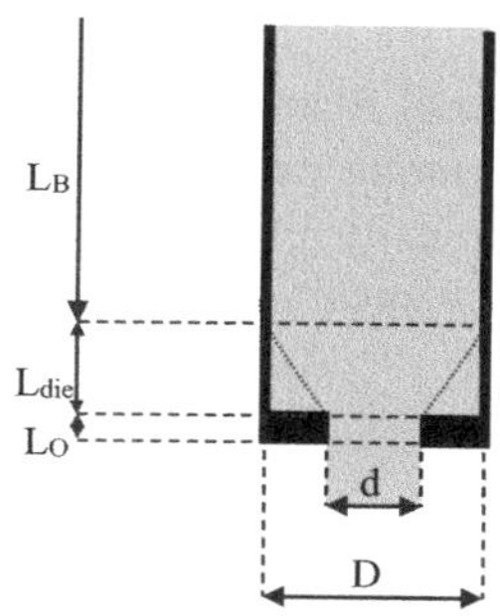

Figure 2.5. *Decomposition of extrusion flow–axisymmetric case. D is the diameter of the piston, d is the diameter of the die, L_{die} is the length of the conical zone, L_O is the length of the final pipe and L_B is the length of the plug flow zone. For a color version of this figure, see www.iste.co.uk/perrot/3dconcrete.zip*

Some studies suggest that extrusion pressure is proportional to the material's shear yield stress (Perrot et al. 2012; Zhou et al. 2013).

The extrusion stage can induce specific problems such as aggregate blocking (El Cheikh et al. 2017) or material consolidation in the constriction (Perrot et al. 2007). These problems are linked to the size and volume fraction of the largest aggregates, which can impart granular-like behavior to the material. It is therefore advisable to select the size of the largest aggregates in the composition according to the size of the die and not to exceed a critical volume fraction of aggregates (around 80% of the maximum compactness of the aggregate mix) to maintain viscoplastic behavior (Yammine et al. 2008).

Regarding the consolidation phenomenon, during the extrusion of cementitious materials, there is competition between the kinetics of interstitial fluid drainage influenced by the pressure gradient in the die, and the speed of material advancement during extrusion (Toutou et al. 2005; Perrot et al. 2009, 2014; Khelifi et al. 2013). The appearance of drainage will increase the material's frictional behavior, heighten the risk of defects and induce a rise in pumping pressure. To counter this phenomenon, printable cementitious materials often contain a viscosifying agent to slow down drainage kinetics by increasing the viscosity of the interstitial fluid, thereby preventing this consolidation phenomenon (Varela et al. 2023).

2.4.4. *Filament deposition and local stability*

Managing the deposition of material filaments to form structure layers is one of the crucial points in defining the printing process. Indeed, the deposition strategy will influence the choice and complexity of the robot, the surface finish of the printed structure and also its mechanical characteristics in the hardened state (Duballet et al. 2017).

Several deposition methods have been developed since the early days of extrusion-based 3D concrete printing. Historically, the first developments of this process at Loughborough University or in Southern California with Professor Koshnevis' team followed diametrically opposed solutions.

The Loughborough University team (Le et al. 2012; Lim et al. 2012) chose to deposit small millimetric filaments of cylindrical cross-section that deform under their own weight. The deformation of these cylinders fills the voids created by the shape of these filaments to create a compact structure. In this case, the final height of the filament is dictated by its shear yield stress. The material is subjected to gravity, which induces a vertical stress at the base of the filament equal to $\rho g \Delta h$ with ρ the density of the material and Δh the height of an elementary layer. Equilibrium is reached when the stress generated becomes equal to the material's elongational shear yield stress.

In contrast, for processes such as Contour Crafting or ConPrint3D in Southern California (Khoshnevis 2004; Nerella and Mechtcherine 2019), the layers have a rectangular cross-section, which is intended to be that of the die. In this second case, the shear yield stress must be higher than the stress generated by gravity.

These two cases can be considered to be asymptotic limits of possible deposition strategies. Indeed, the Loughborough process corresponds to a free deposition that is dictated solely by the effect of gravity on the filament leaving the printhead, whereas the contour crafting process uses the so-called "infinite brick" process, where the shape of the bead is influenced solely by extrusion and deformation of the bead in the die.

It should be noted that free deposition of a material filament can generate regular filament patterns such as alternating loops or a repetition of translating loops that are due to local buckling of the material under its own weight. Geffrault's work (2022; Geffrault et al. 2023) provides a theoretical framework for predicting the pattern of the deposited filament as a function of two scaled parameters: scaled height (ratio of printhead height to die diameter) and scaled speed (ratio of robot feed speed to filament exit speed). This analysis makes it possible to design complex multi-outlet dies that can produce an optimized wall structure in a single pass, without the need for additional mechanized systems (Geffrault 2022).

In all cases, to manage the deposition process properly, it is necessary to adjust the extrudate exit speed to the printhead translation speed, so as not to generate tensile stress that could lead to filament breakage, or excessive compressive stress in the filament if the extrusion rate is too high (Carneau et al. 2022).

Between these two processes, it is possible to imagine a whole range of deposition strategies where the shape of the deposited filament is influenced by both gravity and the shape and position of the die.

One of the most widely used solutions is called controlled compression deposition, in which the die imposes a force perpendicular to the filament deposition plane and fixes the height of the deposited layer. The rheology of the material and its evolution over time, as well as the force imposed by the die, must ensure that the pressure exerted does not deform the support layers deposited underneath.

In this work, Carneau et al. (2022) were able to theoretically describe the various controlled compression deposition processes by analyzing the deposition conditions in the same frame of reference as Geffrault et al. (2023) as a function of the geometric and kinematic parameters of the process and the rheological behavior of the cementitious material (Figure 2.10).

Free deposition under gravity is observed when the gravity stress induced by a layer height equal to the deposition height (between the extrusion nozzle and the deposition surface) is greater than the material's elongational shear yield stress. The material cannot support the load and flows under its own weight to an equilibrium height. It is important to note that this strategy can lead to poor control of the geometry of the printed structure, as the deformations of each layer can accumulate and result in significant print defects. Carneau uses the dimensionless weight w^* to describe the occurrence of this deposition mode.

$$w^* = \frac{\rho g H}{\tau_0\sqrt{3}} > 1 \text{ or } H^* > \frac{\tau_0\sqrt{3}}{\rho g D} \qquad [2.13]$$

Examination of the scaled velocity indicates whether tensile fractures will occur during deposition. The limit of this phenomenon, known as under-extrusion, can also be plotted on the (H*, V*) diagram.

Finally, a comparison of the forces acting on the deposited layer should enable us to check whether the sum of the force exerted by the die is sufficient to deform the deposited layer, but not so great as to impact the layers deposited underneath. A first limit is established through a kinematic analysis: this upper limit for controlled compression of the layer depends on ensuring that the material has time to reach the support layer before the print head completely passes its position. This condition is written as follows in the dimensioned reference frame (H*,V*): H*.V* <1.

The other limit is linked to the strength of the lower support layer, which receives the compressed layer deposited by the nozzle. This resistance also depends on the A_{thix} structural build-up rate (the rate of increase in the shear yield stress in Pa/s of the material left at rest) of the cementitious material. Carneau et al. (2022) define a dimensionless structural build-up velocity A*, which is used to define the lower limit of the controlled deposition strategy in the (H*,V*) reference frame by balancing the forces on the deposition support layer (equation [2.14]).

$$2w^* + \tau^* - A^* < 1 \text{ with } A^* = \frac{A_{thix}\Delta t}{\tau_0} \qquad [2.14]$$

Here, Δt represents the time between two consecutive deposits and τ* being the characteristic compressive stress of a layer, which can be written as follows:

$$\tau^* = \frac{4F}{\pi D^2 \tau_0 \sqrt{3}} \qquad [2.15]$$

Carneau suggests using the compression work of a plastic fluid to write the compressive stress of a layer of material between two surfaces (Coussot 2005). The compressive stress of a layer can thus be written as:

$$F = \frac{\tau_0 D B^2}{2H} \qquad [2.16]$$

Here, B is the width of the deposited filament. The analysis proposed by Carneau allows us to predict the shape of the deposited filaments as a function of the geometric and kinematic parameters and the rheological behavior of the material, as shown in Figure 2.6.

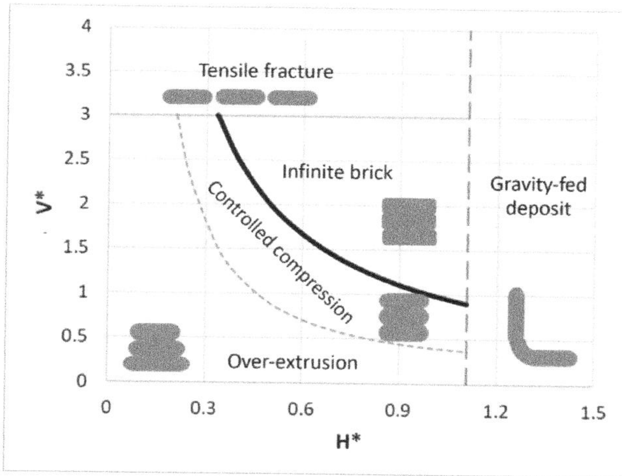

Figure 2.6. *Cementitious material filament deposition strategies and situations in extrusion-deposition 3D printing from Carneau et al. (2022). For a color version of this figure, see www.iste.co.uk/perrot/3dconcrete.zip*

Finally, in the case of cantilever structure printing using a three- or four-axis robot, it will be necessary to check that the deformation remains elastic for the part of the material not supported by an underlying layer (i.e. in the void). A check on the stability of the cantilever can be carried out to ensure that the material deflection at the end of the cantilever is not too great. This can be done by adopting a cantilever beam hypothesis, or by numerical simulation using finite elements, for example (Figure 2.11).

The use of printed temporary supports is also an interesting alternative in this case, as in the case of fuse deposition modeling techniques for plastics.

2.4.5. *Overall structure stability during printing*

2.4.5.1. *Fracture mechanisms of printed structures*

The issue of stability of a concrete structure during the manufacturing process is undoubtedly the most studied topic in the field of 3D printing by extrusion and filament deposition (Mechtcherine et al. 2020).

Researchers have thus defined the ability of a cementitious material to support the weight of the structure being printed with the terms "buildability" or "printability" (Le et al. 2012), which translates the material's ability to be printed and to support the weight of the layers deposited on top of it.

The analysis of the problem identified two modes of failure for structures undergoing printing. The first mode is related to compressive failure of the base layer, which receives the heaviest loading (Perrot et al. 2016; Wangler et al. 2016). For this mode, failure occurs when the stress due to the weight of the printed layers exceeds the compressive elongational yield stress of the base layer material. This failure is the mode generally encountered for low-slung structural geometries (Roussel 2018).

The second reported mode of failure is that associated with buckling of the printed structure linked to a lack of material rigidity (Wolfs et al. 2018; Suiker et al. 2020). This failure occurs when printing slender or complex structures with overhangs.

2.4.5.2. *Compression failure of the base layer*

Compressive failure of the base layer is a problem described by Perrot et al. (2016). This mode of failure occurs if the material's compressive strength $\sigma_0(t)$ becomes less than the load acting on the first printed layer. This load is directly related to the weight deposited on the layer and therefore to the increasing height $h(t)$ of the structure being printed.

To describe the failure, we need to compare the temporal evolution of the mechanical strength of the cementitious material and the mechanical load due to the elevation of the printed form. The theory developed, on the one hand, indicates whether the layered structure is capable of supporting its own weight and, on the other, predicts when the structure will collapse due to the compressive failure of the first layer.

In the case of the construction of a vertical wall or column, the vertical stress acting on the first layer deposited increases with time and with the elevation of the structure. In their work, Perrot et al. (2016) consider a constant average printing speed over the manufacturing time. This speed corresponds to the average vertical elevation speed of the structure and is denoted by R. It is important to note that the print elevation speed R depends on several factors such as the length of the contour to be printed L_c and the extruder feed speed V_e (Roussel 2018). In this case, the elevation speed depends on the geometry and can be written as $R = V_e H / L_c$, where H is the thickness of a layer. The shorter the length of the contour to be printed, the faster the elevation speed.

The elevation speed can then be used to write the evolution of the vertical stress σ_v acting on the first layer as a function of time and the density of the cementitious material ρ:

$$\sigma_V = \rho g h(t) = \rho g R t \qquad [2.17]$$

The stability of the first layer and therefore of the structure can be assessed by comparing this vertical stress with the compressive strength of the material. This resistance, which according to Perrot et al. (2016), drawing on work on the squeeze test (Roussel and Lanos 2003), depends on the geometry of the printed layer, can be determined from the following relationship:

$$\sigma_c(t) = \alpha_{geom} \cdot \tau_0(t) \qquad [2.18]$$

where α_{geom} is a geometric factor depending on the cross-section of the deposited layers. Accurate estimation of material structural build-up over time to obtain the evolution of the shear yield stress over time $\tau_0(t)$ (more commonly measured than the elongation one) is necessary to be able to compare the loading and strength of the first layer.

In the simplest case, it is possible to use a constant A_{thix} structural build-up rate, as proposed by Roussel et al. (2006). In this case, the stability of the structure is controlled by the A_{thix}/R ratio, and it is possible to determine the instant of compressive failure of the first layer:

$$t_f = \frac{\tau_{0,0}}{\rho g R / \alpha_{geom} - A_{thix}} \qquad [2.19]$$

However, the linear description of the evolution of the shear yield stress may prove too simplistic to predict the effective strength of the first layer. In their work, Perrot et al. (2016) showed that the use of expression [2.19] could lead to the calculation of an erroneous failure time when the shear yield stress does not follow a linear evolution.

Also, the appearance of pressure-dependent behavior with a material presenting a nonzero angle of friction can favor the compressive strength of the first layer deposited.

2.4.5.3. *Buckling failure*

Another reported cause of structure failure during printing is related to elastic buckling of the structure. This mode of failure was initially described by researchers at the Technical University of Eindhoven (Wolfs et al. 2018). Figure 2.7 illustrates the buckling collapse of a printed square-section structure. This problem can be described theoretically by studying the self-weight buckling problem of a structure with a vertical elastic stiffness gradient (Suiker et al. 2020). This mode of failure will drive stability during the printing of slender, complex structures.

In this case, in addition to the accuracy with which the layers are laid, which necessarily leads to eccentricities resulting in buckling instability, it is the evolution of the elastic modulus with rest time that will drive the instability of the structure under its own weight during printing.

Figure 2.7. *Structure undergoing buckling during printing*

The use of Euler's buckling theory indicates that in the case of a strictly vertical element, the buckling limit height evolves with the cube root of the material's elastic stiffness. Thus, considering a constant elastic modulus over the entire height of the structure, it is possible to propose the following relationship to calculate the critical height h_c leading to buckling of the structure (Roussel 2018):

$$h_c = \left(\frac{8EI}{\rho g A}\right)^{1/3} \qquad [2.20]$$

with I the squared moment of the printed layer and A its cross-section in the horizontal plane. Equation [2.20] demonstrates the importance of layer design in minimizing the risk of buckling.

A more precise analysis of the problem can be obtained by taking into account the elastic stiffness gradient along the height of the printed structure. An analytical development has been proposed by Suiker et al. (2020) to predict the buckling stability of vertical structures during printing.

For more complex structures with cantilevers, numerical simulation is an appropriate solution to ensure stability during printing. However, it is important to select the right material behavior laws (and their temporal evolution) and accurately reproduce printing defects, which can often be sources of buckling initiation (Perrot et al. 2021).

2.4.6. *Elastic deformation and printing precision*

In addition to verifying the stability of the structure, we also need to ensure the fidelity of the print by checking the deformations of each layer under loading (each layer having, for a given time, its own rigidity due to different resting times). Depending on the type of deposition strategy chosen, the impact of elastic deformation is not the same: for free deposition under gravity, the deformations accumulate and can lead to a significant discrepancy between the expected height and the height obtained, or even to buckling during filament deposition. In contrast, with the controlled compression deposition strategy, each nozzle passage corrects these deformations, and the height of the structure is fixed by the passage of the printing nozzle.

In the case of cumulative deformations, it is possible to estimate the elastic deformation of the first layer deposited during printing. Values of the initial elastic moduli of printed cementitious materials are of the order of 100 kPa (Perrot et al. 2016; Wolfs et al. 2018). In addition, elastic moduli evolve in the same way as shear yield stresses and it is possible to consider a linear evolution of the elastic modulus as proposed by Roussel.

The resulting deformations can be non-negligible, demonstrating the need to take this phenomenon into account by ensuring sufficient stiffening speeds or reducing the elevation speed of the structure.

2.4.7. *Material curing during and after printing*

2.4.7.1. *Tensile fracture cracking*

During the impression of a curve, in a bend, the inside part of the bend is compressed, while the outside fiber is stretched. The risk of cracking increases as the radius of curvature decreases and the trajectory changes direction abruptly. The elongation undergone by the material can then be greater than the critical deformation of the material and create a crack that is detrimental to the aesthetics of the printed part, but above all, detrimental to the strength and durability of the hardened material (Perrot and Amziane 2019). This cracking is also detrimental for special applications such as underwater 3D printing (Mazhoud et al. 2019; Jacquet et al. 2021b).

Solutions to increase resistance to cracking include the use of fibers or the addition of viscosity modifying agents. These solutions make it possible to increase tensile strength while increasing strain at failure.

It will therefore be necessary to check that, when printing curved structures, the local radii of curvature do not generate deformation greater than the material's critical deformation, which can be determined experimentally by console beam tests or tensile tests on fresh materials (Jacquet et al. 2020; Jacquet et al. 2021a).

2.4.7.2. *Drying and induced shrinkage*

Unlike concrete in a formwork, printed structures are not protected at an early age and are therefore directly exposed to their environment. Interaction and exchange with the direct environment can lead to drying in dry weather, or even washout of the material during wet weather on site.

During drying, the evaporation of water produces an exchange of a quantity of water toward the outside, the rate of which depends on the direct environment (temperature, humidity and wind speed). The water on the surface is the first to evaporate, creating a pressure gradient dominated by capillary pressure, which tends to transport water from the core of the material to the surface to rebalance the water content (Keita et al. 2019). This feed flow can be slowed down by reducing the permeability of the granular skeleton (grain size reduction) or by increasing the viscosity of the pore fluid (addition of viscosity modifying agents).

The use of curing products or blankets is also a solution for reducing exchanges with the outside world, bringing the situation closer to that of formwork concrete.

These precautions are important, as drying can lead to deleterious phenomena: significant dimensional shrinkage, lack of water for the hydration reaction, surface cracking and spalling.

Taking into account the direct environment (temperature, wind speed, etc.) is therefore essential (in the same way as for the casting of large-area slabs) to take into account any drying that may be detrimental to the mechanical quality of the printed structure.

2.4.7.3. *Bonding between layers – weakness at the interface between layers*

One of the most frequently reported shortcomings of concrete 3D printing by extrusion and filament deposition concerns interface quality and interlayer bonding.

Indeed, an interface between layers of good mechanical quality will confer good resistance to the material once hardened. Several solutions have been proposed in the literature to achieve this.

The first is based on work on multi-layer casting in self-placing concretes (Roussel and Cussigh 2008; Wangler et al. 2016) and consists in forcing a mixture of the material in the two layers by attempting to destructure the material in the underlying layer.

A second solution is to avoid or limit drying of the layer already deposited, as shown by the work of Keita et al. (2019). For example, the use of super-absorbent polymers (Nerella et al. 2017) and viscosity modifying agents is recommended to reduce the material's permeability and/or susceptibility to drying. Drying can also be reduced by wetting the underlying layer before deposition, which has been shown to have a positive effect on the mechanical behavior of printed structures (Sanjayan et al. 2018).

The third solution involves adding transverse reinforcements at the interfaces (such as nails, screws or staples) (Hass and Bos 2020; Perrot et al. 2020) or a bonding agent to ensure good mechanical properties of the interlayer and avoid possible delamination.

Finally, it is possible to increase the contact surface between layers by creating grooves or striation on the surface of the layers. In this way, the underside of the deposited layer becomes embedded in the pattern of the support layer, increasing the contact area and hence the mechanical strength of the interface. Indeed, the greater the surface area of the layers in contact, the greater the strength of the interface (Van Der Putten et al. 2019).

In all cases, the quality of the interface is crucial to the mechanical behavior of the printed material. It is important to note that good surface quality will result in a homogeneous material behaving like a monolithic structure, whereas poor quality will result in a laminated material sensitive to delamination.

In all cases, it is often necessary to limit the time between deposits with a maximum value t_{max} to guarantee a good quality interface by reducing drying and facilitating mixing.

2.5. Conclusion

In this chapter, we presented various methods of 3D printing by extrusion and filament deposition, highlighting the diversity of situations and applications reported in the literature and used by the construction industry.

After presenting the fresh-state behavior of cementitious materials, in flow and at rest, we linked the main stages of the extrusion printing process by describing the behavioral parameters involved.

While rheological behavior is involved in material transport and shaping, resting behavior (after strand deposition), characterized by hardening and stiffening kinetics, will limit the speed of the process to ensure the overall stability of the structure throughout printing.

Also, the precision of placement and the elastic modulus of the material play a part in buckling stability and the feasibility of cantilever printing.

Finally, we emphasized the importance of curing the material after printing to limit water exchange and guarantee optimum mechanical performance in the hardened state.

2.6. References

Adams, M.J., Aydin, I., Briscoe, B.J., Sinha, S.K. (1997). A finite element analysis of the squeeze flow of an elasto-viscoplastic paste material. *Journal of Non-Newtonian Fluid Mechanics*, 71(1–2), 41–57.

Asprone, D., Auricchio, F., Menna, C., Mercuri, V. (2018). 3D printing of reinforced concrete elements: Technology and design approach. *Construction and Building Materials*, 165, 218–231.

Benbow, J. and Bridgwater, J. (1993). *Paste Flow and Extrusion*. Clarendon Press, Oxford.

Carneau, P., Mesnil, R., Roussel, N., Baverel, O. (2020). Additive manufacturing of cantilever – From masonry to concrete 3D printing. *Automation in Construction*, 116, 103184.

Carneau, P., Mesnil, R., Baverel, O., Roussel, N. (2022). Layer pressing in concrete extrusion-based 3D-printing: Experiments and analysis. *Cement and Concrete Research*, 155, 106741.

Chen, H., Zhang, D., Chen, P., Li, N., Perrot, A. (2023). A review of the extruder system design for large-scale extrusion-based 3D concrete printing. *Materials*, 16(7), 2661.

Choi, M.S., Kim, Y.J., Kim, J.K. (2014). Prediction of concrete pumping using various rheological models. *International Journal of Concrete Structures and Materials*, 8, 269–278.

Chopin, D., de Larrard, F., Cazacliu, B. (2004). Why do HPC and SCC require a longer mixing time? *Cement and Concrete Research*, 34(12), 2237–2243.

Coussot, P. (2005). *Rheometry of Pastes, Suspensions, and Granular Materials: Applications in Industry and Environment*. John Wiley & Sons, New York.

De Schryver, R., El Cheikh, K., Lesage, K., Yardimci, M.Y., De Schutter, G. (2021). Numerical reliability study based on rheological input for Bingham paste pumping using a finite volume approach in OpenFOAM. *Materials*, 14(17), 1–23.

De Schutter, G. and Feys, D. (2016). Pumping of fresh concrete: Insights and challenges. *RILEM Technical Letters*, 1, 76–80.

Dils, J., De Schutter, G., Boel, V. (2012). Influence of mixing procedure and mixer type on fresh and hardened properties of concrete: A review. *Materials and Structures*, 45, 1673–1683.

Duballet, R., Baverel, O., Dirrenberger, J. (2017). Classification of building systems for concrete 3D printing. *Automation in Construction*, 83, 247–258.

El Cheikh, K., Rémond, S., Khalil, N., Aouad, G. (2017). Numerical and experimental studies of aggregate blocking in mortar extrusion. *Construction and Building Materials*, 145, 452–463.

Esposito, L., Casagrande, L., Menna, C., Asprone, D., Auricchio, F. (2021). Early-age creep behaviour of 3D printable mortars: Experimental characterisation and analytical modelling. *Materials and Structures*, 54, 1–16.

Feys, D., De Schutter, G., Fataei, S., Martys, N.S., Mechtcherine, V. (2022). Pumping of concrete: Understanding a common placement method with lots of challenges. *Cement and Concrete Research*, 154, 106720.

Gaudillière, N., Dirrenberger, J., Duballet, R., Bouyssou, C., Mallet, A., Roux, P., Zakeri, M., Xtree, E.R. (2020). Industrialising concrete 3D printing: Three case studies. In *Design Transactions: Rethinking Information Modelling for a New Material Age*, Sheil, B., Thomsen, M.R., Tamke, M., Hanna, S. (eds). UCL Press, London.

Geffrault, A. (2022). *Mécanismes d'écoulement et forme des dépôts dans l'impression 3D en béton*. Écoles des Ponts ParisTech, Paris.

Geffrault, A., Bessaies-Bey, H., Roussel, N., Coussot, P. (2023). Printing by yield stress fluid shaping. *Additive Manufacturing*, 103752.

Gosselin, C., Duballet, R., Roux, P., Gaudillière, N., Dirrenberger, J., Morel, P. (2016). Large-scale 3D printing of ultra-high performance concrete – A new processing route for architects and builders. *Materials & Design*, 100, 102–109.

Hass, L. and Bos, F. (2020). Bending and pull-out tests on a novel screw type reinforcement for extrusion-based 3D printed concrete. In *Second RILEM International Conference on Concrete and Digital Fabrication: Digital Concrete 2020*. Springer, Cham.

Jacquet, Y., Picandet, V., Rangeard, D., Perrot, A. (2020). Gravity induced flow to characterize rheological properties of printable cement-based materials. *RILEM Technical Letters*, 5, 150–156.

Jacquet, Y., Perrot, A., Picandet, V. (2021a). Assessment of asymmetrical rheological behavior of cementitious material for 3D printing application. *Cement and Concrete Research*, 140, 106305.

Jacquet, Y., Picandet, V., Perrot, A. (2021b). Characterization of tensile behavior of fresh cementitious materials. *ACI Materials Journal*, 118(6), 217–226.

Jézéquel, P.-H. and Collin, V. (2007). Mixing of concrete or mortars: Dispersive aspects. *Cement and Concrete Research*, 37(9), 1321–1333.

Kaplan, D. (1999). Pompage des bétons. PhD Thesis, École des ponts ParisTech, Paris.

Kaplan, D., de Larrard, F., Sedran, T. (2005). Design of concrete pumping circuit. *ACI Materials Journal*, 102(2), 110.

Keita, E., Bessaies-Bey, H., Zuo, W., Belin, P., Roussel, N. (2019). Weak bond strength between successive layers in extrusion-based additive manufacturing: Measurement and physical origin. *Cement and Concrete Research*, 123, 105787.

Khelifi, H., Perrot, A., Lecompte, T., Rangeard, D., Ausias, G. (2013). Prediction of extrusion load and liquid phase filtration during ram extrusion of high solid volume fraction pastes. *Powder Technology*, 249, 258–268.

Khoshnevis, B. (2004). Automated construction by contour crafting – Related robotics and information technologies. *Automation in Construction*, 13(1), 5–19.

Khoshnevis, B., Bodiford, M., Burks, K., Ethridge, E., Tucker, D., Kim, W., Toutanji, H., Fiske, M. (2005). Lunar contour crafting – A novel technique for ISRU-based habitat development. In *43rd AIAA Aerospace Sciences Meeting and Exhibit*. NASA Technical Reports Server (NTRS), Hampton, Virginia.

Kruger, J., Zeranka, S., van Zijl, G. (2019). An ab initio approach for thixotropy characterisation of (nanoparticle-infused) 3D printable concrete. *Construction and Building Materials*, 224, 372–386.

Kwon, S.H., Jang, K.P., Kim, J.H., Shah, S.P. (2016). State of the art on prediction of concrete pumping. *International Journal of Concrete Structures and Materials*, 10, 75–85.

Le, T.T., Austin, S.A., Lim, S., Buswell, R.A., Gibb, A.G., Thorpe, T. (2012). Mix design and fresh properties for high-performance printing concrete. *Materials and Structures*, 45, 1221–1232.

Leach, N., Carlson, A., Khoshnevis, B., Thangavelu, M. (2012). Robotic construction by contour crafting: The case of lunar construction. *International Journal of Architectural Computing*, 10(3), 423–438.

Lecompte, T. and Perrot, A. (2017). Non-linear modeling of yield stress increase due to SCC structural build-up at rest. *Cement and Concrete Research*, 92, 92–97.

Lim, S., Buswell, R.A., Le, T.T., Austin, S.A., Gibb, A.G., Thorpe, T. (2012). Developments in construction-scale additive manufacturing processes. *Automation in Construction*, 21, 262–268.

Mantellato, S., Palacios, M., Flatt, R.J. (2019). Relating early hydration, specific surface and flow loss of cement pastes. *Materials and Structures*, 52(1), 5.

Marchon, D., Kawashima, S., Bessaies-Bey, H., Mantellato, S., Ng, S. (2018). Hydration and rheology control of concrete for digital fabrication: Potential admixtures and cement chemistry. *Cement and Concrete Research*, 112, 96–110.

Massoussi, N., Keita, E., Roussel, N. (2017). The heterogeneous nature of bleeding in cement pastes. *Cement and Concrete Research*, 95, 108–116.

Mazhoud, B., Perrot, A., Picandet, V., Rangeard, D., Courteille, E. (2019). Underwater 3D printing of cement-based mortar. *Construction and Building Materials*, 214, 458–467.

Mechtcherine, V., Bos, F.P., Perrot, A., Da Silva, W.L., Nerella, V.N., Fataei, S., Wolfs, R.J., Sonebi, M., Roussel, N. (2020). Extrusion-based additive manufacturing with cement-based materials – Production steps, processes, and their underlying physics: A review. *Cement and Concrete Research*, 132, 106037.

Mettler, L.K., Wittel, F.K., Flatt, R.J., Herrmann, H.J. (2016). Evolution of strength and failure of SCC during early hydration. *Cement and Concrete Research*, 89, 288–296.

Nerella, V.N. and Mechtcherine, V. (2019). Studying the printability of fresh concrete for formwork-free concrete onsite 3D printing technology (CONPrint3D). In *3D Concrete Printing Technology*, Sanjayan, J.G., Nazari, A., Nematollahi, B. (eds). Butterworth–Heinemann, Oxford [Online]. Available at: https://www.sciencedirect.com/science/article/pii/B9780128154816000166.

Nerella, V.N., Hempel, S., Mechtcherine, V. (2017). Micro- and macroscopic investigations on the interface between layers of 3D-printed cementitious elements. In *Proceedings of the International Conference on Advances in Construction Materials and Systems*, Chennai.

Nicolas, R., Richard, B., Nicolas, D., Irina, I., Temitope, K.J., Dirk, L., Viktor, M., Romain, M., Arnaud, P., Ursula, P. (2022). Assessing the fresh properties of printable cement-based materials: High potential tests for quality control. *Cement and Concrete Research*, 158, 106836.

Ovarlez, G. and Roussel, N. (2006). A physical model for the prediction of lateral stress exerted by self-compacting concrete on formwork. *Materials and Structures*, 39, 269–279.

Perrot, A. and Amziane, S. (2019). 3D printing in concrete: General considerations and technologies. In *3D Printing of Concrete: State of the Art and Challenges of the Digital Construction Revolution*, Perrot, A. (ed.). ISTE Ltd, London, and John Wiley & Sons, New York.

Perrot, A. and Rangeard, D. (2016). Effects of mix design parameters on consolidation behaviour of fresh cement-based materials. *Materials and Structures*, 50(2), 117.

Perrot, A., Lanos, C., Melinge, Y., Estellé, P. (2007). Mortar physical properties evolution in extrusion flow. *Rheologica Acta*, 46, 1065–1073.

Perrot, A., Rangeard, D., Melinge, Y., Estelle, P., Lanos, C. (2009). Extrusion criterion for firm cement-based materials. *Applied Rheology*, 19(5), 53042-1–53042-7.

Perrot, A., Mélinge, Y., Rangeard, D., Micaelli, F., Estellé, P., Lanos, C. (2012). Use of ram extruder as a combined rheo-tribometer to study the behaviour of high yield stress fluids at low strain rate. *Rheologica Acta*, 51, 743–754.

Perrot, A., Pierre, A., Vitaloni, S., Picandet, V. (2015). Prediction of lateral form pressure exerted by concrete at low casting rates. *Materials and Structures*, 48(7), 2315–2322. doi: 10.1617/s11527-014-0313-8.

Perrot, A., Rangeard, D., Pierre, A. (2016). Structural built-up of cement-based materials used for 3D-printing extrusion techniques. *Materials and Structures*, 49, 1213–1220.

Perrot, A., Rangeard, D., Nerella, V.N., Mechtcherine, V. (2019). Extrusion of cement-based materials – An overview. *RILEM Technical Letters*, 3, 91–97.

Perrot, A., Jacquet, Y., Rangeard, D., Courteille, E., Sonebi, M. (2020). Nailing of layers: A promising way to reinforce concrete 3D printing structures. *Materials*, 13(7), 1518.

Perrot, A., Pierre, A., Nerella, V.N., Wolfs, R.J.M., Keita, E., Nair, S.A.O., Neithalath, N., Roussel, N., Mechtcherine, V. (2021). From analytical methods to numerical simulations: A process engineering toolbox for 3D concrete printing. *Cement and Concrete Composites*, 122, 104164.

Reiter, L., Wangler, T., Anton, A., Flatt, R.J. (2020). Setting on demand for digital concrete – Principles, measurements, chemistry, validation. *Cement and Concrete Research*, 132, 106047.

Roussel, N. (2005). Steady and transient flow behaviour of fresh cement pastes. *Cement and Concrete Research*, 35(9), 1656–1664.

Roussel, N. (2006). A thixotropy model for fresh fluid concretes: Theory, validation and applications. *Cement and Concrete Research*, 36(10), 1797–1806. doi: 10.1016/j.cemconres.2006.05.025.

Roussel, N. (2018). Rheological requirements for printable concretes. *Cement and Concrete Research*, 112, 76–85.

Roussel, N. and Cussigh, F. (2008). Distinct-layer casting of SCC: The mechanical consequences of thixotropy. *Cement and Concrete Research*, 38(5), 624–632.

Roussel, N. and Lanos, C. (2003). Plastic fluid flow parameters identification using a simple squeezing test. *Applied Rheology*, 13(3), 132–141.

Roussel, N., Lemaître, A., Flatt, R.J., Coussot, P. (2010). Steady state flow of cement suspensions: A micromechanical state of the art. *Cement and Concrete Research*, 40(1), 77–84.

Roussel, N., Ovarlez, G., Garrault, S., Brumaud, C. (2012). The origins of thixotropy of fresh cement pastes. *Cement and Concrete Research*, 42(1), 148–157.

Roussel, N., Bessaies-Bey, H., Kawashima, S., Marchon, D., Vasilic, K., Wolfs, R. (2019). Recent advances on yield stress and elasticity of fresh cement-based materials. *Cement and Concrete Research*, 124, 105798.

Sanjayan, J.G., Nematollahi, B., Xia, M., Marchment, T. (2018). Effect of surface moisture on inter-layer strength of 3D printed concrete. *Construction and Building Materials*, 172, 468–475.

Schack, R., Krause, M., Näther, M., Nerella, V.N. (2017). CONPrint3D: 3D-concrete-printing as an alternative for masonry. *Bauingenieur*, 92, 355–363.

Secrieru, E., Khodor, J., Schröfl, C., Mechtcherine, V. (2018). Formation of lubricating layer and flow type during pumping of cement-based materials. *Construction and Building Materials*, 178, 507–517.

Sleiman, H., Perrot, A., Amziane, S. (2010). A new look at the measurement of cementitious paste setting by Vicat test. *Cement and Concrete Research*, 40(5), 681–686.

Souza, M.T., Ferreira, I.M., de Moraes, E.G., Senff, L., de Oliveira, A.P.N. (2020). 3D printed concrete for large-scale buildings: An overview of rheology, printing parameters, chemical admixtures, reinforcements, and economic and environmental prospects. *Journal of Building Engineering*, 32, 101833.

Subramaniam, K.V. and Wang, X. (2010). An investigation of microstructure evolution in cement paste through setting using ultrasonic and rheological measurements. *Cement and Concrete Research*, 40(1), 33–44. doi: 10.1016/j.cemconres.2009.09.018.

Suiker, A.S., Wolfs, R.J., Lucas, S.M., Salet, T.A. (2020). Elastic buckling and plastic collapse during 3D concrete printing. *Cement and Concrete Research*, 135, 106016.

Tao, Y., Rahul, A.V., Lesage, K., Van Tittelboom, K., Yuan, Y., De Schutter, G. (2022). Mechanical and microstructural properties of 3D printable concrete in the context of the twin-pipe pumping strategy. *Cement and Concrete Composites*, 125, 104324.

Tattersall, G.H. (1976). *The Workability of Concrete*. Portland Cement Association, Washington, DC.

Tattersall, G.H. and Banfill, P.F.G. (1983). *The Rheology of Fresh Concrete*. Pitman Advanced Publishing Program, London.

Tavangar, T., Hosseinpoor, M., Yahia, A., Khayat, K.H. (2022). Novel tri-viscous model to simulate pumping of flowable concrete through characterization of lubrication layer and plug zones. *Cement and Concrete Composites*, 126, 104370.

Toutou, Z., Roussel, N., Lanos, C. (2005). The squeezing test: A tool to identify firm cement-based material's rheological behaviour and evaluate their extrusion ability. *Cement and Concrete Research*, 35(10), 1891–1899.

Van Der Putten, J., De Schutter, G., Van Tittelboom, K. (2019). Surface modification as a technique to improve inter-layer bonding strength in 3D printed cementitious materials. *RILEM Technical Letters*, 4, 33–38.

Vandenberg, A., Bessaies-Bey, H., Wille, K., Roussel, N. (2019). Enhancing printable concrete thixotropy by high shear mixing. In *First RILEM International Conference on Concrete and Digital Fabrication – Digital Concrete 2018*. Springer, Cham.

Varela, H., Barluenga, G., Perrot, A. (2023). Extrusion and structural build-up of 3D printing cement pastes with fly ash, nanoclays and VMAs. *Cement and Concrete Composites*, 142, 105217.

Wangler, T., Lloret, E., Reiter, L., Hack, N., Gramazio, F., Kohler, M., Bernhard, M., Dillenburger, B., Buchli, J., Roussel, N. (2016). Digital concrete: Opportunities and challenges. *RILEM Technical Letters*, 1, 67–75.

Wangler, T., Scotto, F., Lloret-Fritschi, E., Flatt, R.J. (2020). Residence time distributions in continuous processing of concrete. In *Rheology and Processing of Construction Materials*, Mechtcherine, V., Khayat, K., Secrieru, E. (eds). Springer, Cham.

Wangler, T., Pileggi, R., Gürel, S., Flatt, R.J. (2022). A chemical process engineering look at digital concrete processes: Critical step design, inline mixing, and scaleup. *Cement and Concrete Research*, 155, 106782.

Wolfs, R.J.M., Bos, F.P., Salet, T.A.M. (2018). Early age mechanical behaviour of 3D printed concrete: Numerical modelling and experimental testing. *Cement and Concrete Research*, 106, 103–116.

Wolfs, R.J.M., Bos, F.P., Salet, T.A.M. (2019). Triaxial compression testing on early age concrete for numerical analysis of 3D concrete printing. *Cement and Concrete Composites*, 104, 103344.

Xiao, J., Ji, G., Zhang, Y., Ma, G., Mechtcherine, V., Pan, J., Wang, L., Ding, T., Duan, Z., Du, S. (2021). Large-scale 3D printing concrete technology: Current status and future opportunities. *Cement and Concrete Composites*, 122, 104115.

Yammine, J., Chaouche, M., Guerinet, M., Moranville, M., Roussel, N. (2008). From ordinary rhelogy concrete to self compacting concrete: A transition between frictional and hydrodynamic interactions. *Cement and Concrete Research*, 38(7), 890–896.

Zhou, X., Li, Z., Fan, M., Chen, H. (2013). Rheology of semi-solid fresh cement pastes and mortars in orifice extrusion. *Cement and Concrete Composites*, 37, 304–311.

3

From Laboratory to Practice: Characterizing Fresh and Cured Printed Materials

Nicolas DUCOULOMBIER[1], Victor DE BONO[1], Fatima Zahraa KACHKOUCH[2], Yohan JACQUET[3] and Arnaud PERROT[2]

[1] XtreeE, Rungis, France

[2] IRDL, Université de Bretagne Sud, Lorient, France

[3] Technische Universität Berlin, Germany

3.1. Introduction

Compared with the cast concrete used in traditional construction methods, 3D concrete printing imposes new specifications in the fresh state (Roussel 2018; Nicolas et al. 2022) and induces a potentially more complex anisotropic mechanical behavior in the hardened state (Asprone et al. 2018; Mechtcherine et al. 2021, 2022; Bos et al. 2022).

These new constraints require the implementation of a comprehensive characterization process throughout the lifecycle of the printed element and material. This process necessitates numerous tests at different phases of the printing project:

– the material formulation phase, particularly to check rheological behavior against specifications;

For a color version of all figures in this chapter, see www.iste.co.uk/perrot/3dconcrete.zip.

3D Concrete Printing,
coordinated by Arnaud PERROT and Yohan JACQUET.

– the printing phase in an industrial context, with the implementation of a quality control system;

– the in-service phase, with the implementation of a mechanical behavior characterization program.

The aim of this chapter is to provide an overview of the characterizations required in these three phases of a printing project.

3.2. Characterization of fresh materials

3.2.1. *Laboratory formulation – specifications*

The formulation of a material for 3D printing involves rheological behavior characterization campaigns to ensure that the material meets all the specifications imposed by the various stages of the printing process. The mixture must be pumpable and possess a certain mechanical strength, characterized by a yield stress called flow and a rigidity characterized by an apparent elastic modulus high enough at an instant t to ensure the stability of the structure during printing (Perrot et al. 2016; Wolfs et al. 2018).

3.2.1.1. *Rheological behavior under steady-state conditions*

To ensure the pumpability of mixes, the rheological behavior under steady-state conditions needs to be described by means of a rheometric test. As explained in Chapter 2, flowing cementitious materials exhibit viscoplastic behavior, which can be described by the Bingham, modified Bingham or Herschel–Bulkley models.

Several flow geometries can be used to analyze the rheological behavior of cementitious materials: parallel-plate geometry or coaxial geometry (Figure 3.1).

Parallel-plate geometry is often dedicated to cement paste, even though some planar rheometers allow concrete testing (necessary representative volume imposed by the size of the largest aggregate), whereas coaxial geometries are available at all scales (Feys et al. 2018; Perrot and Amziane 2024; Yahia and Perrot 2024).

The test protocol must allow the material to be destructured to erase its history of stress and material structuring. Tests generally start with shearing at high speed, which destructures the material. Then, stepwise ramps of rotational speeds are imposed, and resistive torques are measured to scan a sufficient range of shear speeds. A number of data analysis methods can then be used to transform the rotational speed–torque data into flow curves, that is, the evolution of shear flow as a function of shear rate. For example, the Reiner–Riwlin method (Feys et al. 2013, 2023, p. 266; Wallevik et al. 2015) or the method developed by Estellé et al. (2008a, 2008b) appear in the literature as effective solutions for obtaining rheograms.

a) b)

Figure 3.1. *Types of test geometry used to study the flow of cementitious materials: (a) parallel-plate; (b) coaxial geometry*

However, it is important to note that rheological flow behavior is not directly related to pumping behavior due to the phenomenon of particle migration under shear (Secrieru et al. 2020). In this case, it is often the behavior of the lubricating layer that dictates the pumping pressure (Secrieru et al. 2018; Tavangar et al. 2022). It is then possible to set up tests directly on the lubricating layer (cement paste and fine sand) or to encourage the creation of a lubricating layer during the test, as with the linear shear test generated by a slip test in an instrumented tube (Mechtcherine et al. 2014; Nerella and Mechtcherine 2018).

3.2.1.2. *Structural build-up of cementitious materials*

The structural build-up of cementitious materials is often studied by measuring "static" shear yield stress after specific rest times. In this case, the age of the material and the aging conditions must be well controlled. Two time references can be used: the age of the material (after water–cement contact) and the resting time (after the sample has been placed in position for the test). For the study of structural build-up, the second time reference is used, which simulates the aging of the material after it has been deposited in layers (Perrot et al. 2016). Knowledge of the evolution of the static shear yield stress as a function of time enables stability conditions during printing to be verified.

The shear yield stress is often measured using a rheometer equipped with a vane geometry (scissometer) that limits the effects of wall slip. The test is traditionally carried out at a constant shear rate, slow enough to render viscous effects negligible. However, the speed must not be too slow to prevent the material from structuring during the test. In this case, the shear yield stress is calculated from the maximum torque recorded, representative of shear failure of the material (Roussel 2005). This test induces progressive destructuring of the material (transient regime) until it begins to flow. An increasing stress ramp can also be applied to the material, and the shear yield stress in this case corresponds to the sudden acceleration in material deformation recorded during the test. Typical curves for these tests are shown in Figure 3.2.

The cone penetration test is also often used to calculate the shear yield stress from the penetration of the material cone in the material at a given point in time (Lootens et al. 2009; Mazhoud et al. 2019). In this case, the shear yield stress is calculated by balancing the force of gravity acting on the cone and the shear forces generated on its surface.

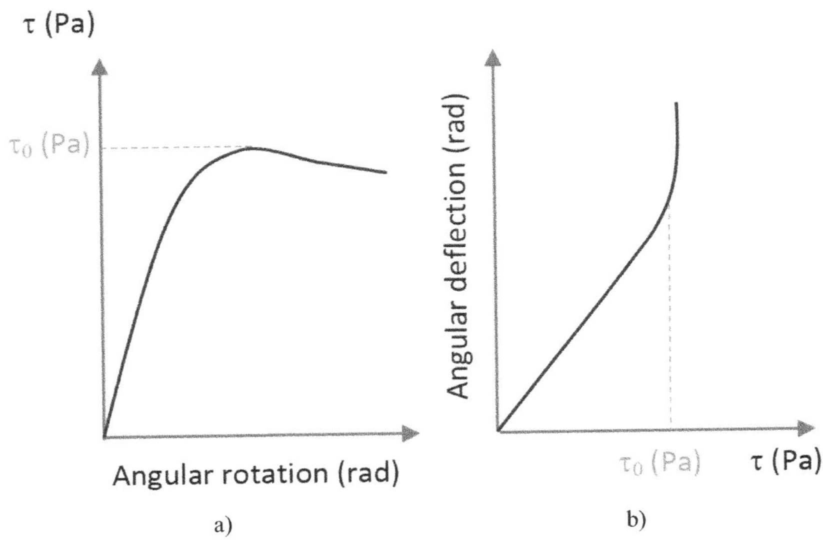

Figure 3.2. *Shear yield stress measurements using Vane geometry: (a) constant speed test; (b) stress ramp*

It is interesting to note that versions of these tests which can be used continuously have been developed. A team at TU Dresden has set up a so-called "single batch" test, in which the same material sample is subjected to successive vane shear tests stopped at peak torque to limit material destructuring (Ivanova and Mechtcherine 2020). This method underestimates the rate of structural build-up by around 10%, but saves considerable time and reduces the dispersion of results caused by the variability of material samples. Similarly, slow-speed penetration tests have been developed to continuously monitor the evolution of the material's shear yield stress (Pott and Stephan 2021; Reiter et al. 2022).

3.2.1.3. *Elongation behavior of materials*

During the printing process, cementitious materials are subjected to compressive stresses related to the support of the layers placed above them. For certain geometries of printed structures, the material may also be in tension (Jacquet et al. 2021a, 2021b). It is therefore interesting to study the material's behavior in compression or tension to more accurately predict the material's stability with regard to the loading of the base layer and the occurrence of cracks in the material's zones, which are under tension.

Regarding compressive behavior, uniaxial compression tests can be used to determine the material's compressive yield point. This yield stress can be measured on materials of different ages in order to monitor the effect of structuring on compressive behavior. When carrying out this type of test, care must be taken to limit wall shear to ensure pure compression loading. It is also important to note that the specimen's slenderness (ratio of height to diameter in the case of a cylindrical specimen) will play a role in the type of loading, and a slenderness of two will be more suitable than a slenderness of one for compressive loading. Finally, optimal specimen dimensions must also be chosen to limit the effects of gravity on the sample and to have a volume adapted to the dimensions of the largest particles.

For tensile behavior, it is possible to test the material under its own weight at the exit of a print head (Jacquet et al. 2020; Ducoulombier et al. 2021), using the so-called "slug test", or to use dumbbell geometries to generate tensile stress in the shrinkage zone. As with compression tests, it is possible to carry out these tests after different resting times to study the evolution of the cementitious material's tensile strength (Jacquet et al. 2021b).

Examples of the geometries used to study the material's behavior in compression and tension are shown in Figure 3.3.

a) b)

Figure 3.3. *Test geometries for studying the elongation behavior of materials: (a) simple compression test; (b) tensile test device*

Finally, it is interesting to note that capillary extrusion tests can be used to study both compressive behavior (elongation failure yield stress in the shaping zone) (Perrot et al. 2012; Varela et al. 2023) and tensile behavior via self-weight failure tests by measuring the mass of filaments (called "slugs" in the literature) at the die exit.

3.2.1.4. *Apparent elastic behavior*

The apparent elastic behavior of cementitious materials in the fresh state is a key parameter for the 3D printing process. Indeed, one of the two modes of failure of structures during printing is related to the elastic buckling of the structure. Printable cementitious materials exhibit significant deformations before material flow, which can be considered at first glance as elastic deformations.

It is first possible to obtain information on the apparent elastic behavior of structures by analyzing simple compression tests. However, care must be taken to use a Hencky strain calculation to account for the large strains obtained in such a test (Jacquet et al. 2022). This methodology provides an apparent material stiffness that can be used pragmatically to ensure that the structure will not buckle during printing. This procedure can also be carried out sequentially to obtain the evolution of the material's apparent elastic modulus over time.

Some authors have suggested using the deformation of a cantilevered beam to estimate the material's elastic modulus in its fresh state (Jacquet et al. 2020). However, this type of measurement is limited to materials that are sufficiently firm for the test to be possible.

Finally, oscillatory rheometry enables a closer examination of the behavior of cementitious materials below the yield stress. Shear flow is often used in this context, and the relationship between the elastic modulus and the shear modulus is applied, assuming a value for the Poisson's ratio. A sinusoidal strain amplitude is imposed on the material at a frequency often close to 1 Hz, and the resulting induced stress amplitude is measured by the rheometer. This method provides continuous access to the loss modulus G'' (viscous part) and the storage modulus G' (elastic part) (Yahia et al. 2023). These moduli are used to calculate the apparent shear modulus G* of the material, which is the ratio of shear stress to strain.

3.2.1.5. *Non-destructive testing*

During the various phases of concrete structure construction, a range of internal pathologies can be encountered. As these are generally undetectable when they first appear, they can adversely affect the quality of the structure. Hence, it has become essential to use non-destructive methods to monitor structures regularly, in depth, and in real time. These techniques also have the advantage of reducing costs and intervention times. In this context, several non-destructive investigation methods have been implemented since the 1960s and are certified by COFREND (Comité Français pour les Essais Non Destructifs).

For 3D concrete printing, Acoustic Emission AE, a technique originating in the 1950s (Drouillard 1994), is undoubtedly the most widely used. The instrumentation and components of the AE apparatus must meet the conditions required by the standards (Fournier 2018). As a passive method, AE can be sensitive to any active process (e.g. stressed materials) and depends in particular on several material-specific factors: when a material is stressed, internal defects (micro-displacements) develop, releasing energy in the form of transient elastic waves. These waves, originating from the material itself, are detected by piezoelectric sensors active in a frequency range extending from 20 kHz to 1 MHz (ultrasonic domain). These sensors transform mechanical waves into electrical signals, which are amplified and read by an acquisition system (Mix 2005; Ohtsu and Grosse 2008), enabling a comprehensive assessment of the internal behavior of materials. Evaluation can be carried out by tracking events in real time, locating a source in 2D or 3D, based on propagation speed, and characterizing the type of source (mechanism produced) by distinguishing the shape of the emitted elastic waves. Standards have been developed to provide certain guidelines specific to the use of AE in concrete (RILEM Technical Committee (Masayasu Ohtsu) 2010; Shen et al. 2020) and to classify damage based on parameters extracted from waveforms, such as the classification of cracks into cracks resulting from tension or other mechanisms (Ohtsu 2006, p. 200; Ohno and Ohtsu 2010; Aggelis 2011). Broadband sensors are recommended for hardened concrete (resonant frequency sensors can also be used with frequencies above 60 kHz) (RILEM Technical Committee (Masayasu Ohtsu) 2010). The choice of frequency is crucial, as it is directly related to the wavelength in the material – the main characteristic of ultrasonic waves.

One of the major advantages of AE is its sensitivity to very low-energy phenomena. The use of AE has therefore been extended to the study of several intrinsic phenomena occurring during concrete setting, which play a crucial role in concrete performance. However, it is important to note the high attenuation of fresh concrete, which is of the order of 2 dB/mm (Iliopoulos et al. 2016), varying according to the frequency used. This implies a signal attenuation of 20 dB for a propagation of 1 cm, which is significant. Drying shrinkage, a phenomenon occurring in relatively fresh concrete, has been studied by several authors (Shiotani et al. 2003; Bacharz et al. 2015). Shiotani et al. (2003) monitored the formation of microcracks due to drying shrinkage in concrete containing glass beads (replacing aggregates) using broadband sensors at 12–960 kHz. They noted the effect of aggregates on AE activities, showing their beneficial effect on shrinkage limitations. Certain AE parameters were used to determine when microcracks develop into macrocracks. McLaskey et al (2007) monitored the formation of microcracks due to drying shrinkage using 200–800 kHz frequency transducers at the surface and depth of the concrete. They established a correlation between the frequency of acoustic emissions and setting time, identifying and classifying shrinkage microcracks by AE. With the same objective of classifying and identifying types of shrinkage microcracks, Bacharz et al. (2015) studied the evolution of microcracks in cementitious paste, on the concrete surface, and at the paste-aggregate interface using 30–80 kHz frequency bandwidth transducers. They noted higher acoustic energy at the concrete surface, compared to other locations.

Some authors have focused on the microstructural activity of concrete. Skal's'kyi et al. (2004) linked acoustic emissions to the development of the crystalline structure, which was almost complete after around six hours, noting a parallel drop in acoustic activity. Van Den Abeele et al. (2009) tested two mortars with different W/C ratios and showed an increase in acoustic emissions for a lower W/C, explained by faster hydration. In particular, they asserted that AE activity begins after temperature peaks, a finding that was also highlighted in Chotard et al. (2001) and Assi et al. (2018). However, other studies (Iliopoulos et al. 2016; Dzaye et al. 2018, 2020) observed high AE activity just after mixing. Further investigation is needed to understand the effect of fresh mixing and the resonant frequency of the sensors used on AE activity to explain the origin of acoustic emissions in the fresh state more deeply. Chotard et al. (2001) linked acoustic parameters (amplitude and frequency) to hydrate formation, using 50–300 kHz wide-bandwidth sensors connected to waveguides. The effect of the W/C ratio observed by Lura et al. (2009) demonstrated the presence of air bubbles by AE, highlighting the phase when mechanical properties begin to develop. Dzaye et al. (2018) studied segregation mechanisms, showing the effect of the density and viscosity of the cementitious paste on the decrease in acoustic energy and propagation velocity. They also demonstrated the sensitivity of AE to the movement of aggregates simulated by falling glass beads.

In the majority of the studies cited above, sensors are placed on waveguides or formwork molds. However, Thirumalaiselvi and Sasmal (2019) conducted their study by placing the sensors directly in contact with the material. This prevented wave propagation for the first two hours after mixing, due to the high attenuation of fresh concrete (Iliopoulos et al. 2016). Nevertheless, and based on the evolution of the acoustic signals, they classified four mechanisms developing in concrete: the formation of CSH, the evaporation of water from the pores, chemical shrinkage and the appearance of microcracks.

Recently, starting in 2022, three studies (Liu et al. 2022; Skibicki et al. n.d.; Yue et al. 2022) have monitored and evaluated the behavior of concretes obtained by 3D printing compared with those cast by the conventional method (using formwork), in the hardened state under mechanical loading. Yue et al. (2022) studied the behavior of printed concretes subjected to splitting and three-point bending, using 35–100 kHz bandpass sensors. They recorded the peak frequencies of the AE, roughly indicating the onset of damage. Peak frequencies were lower for printed concrete than for cast concrete, and higher for tensile cracking modes than for shear cracking modes for both types of concrete. In particular, they integrated the acoustic parameters AF (average frequency) and RA (the ratio between wave propagation time and amplitude), collected during fracture, into an automatic learning algorithm. This allowed them to classify and distinguish between tensile and shear microcracking modes for each type of concrete. Following this calculation based on AE parameters (Yue et al. 2022), it was shown that the percentage of micro-shear cracks in printed concrete is higher than in formwork concrete. Liu et al. (2022) studied concrete reinforced with mesh placed as an internal core in the concrete.

They printed five different lattice structures to evaluate the mechanical properties of concrete under three-point bending using AE and DIC (digital image correlation). A positive correlation was found between acoustic parameters, notably number of hits and energy, and the deformation process during the first three phases of concrete failure. The dominant frequency domain also provided information on internal damage. Low-frequency values (around 100 kHz) were caused by cracking of the hardened cementitious paste, and high-frequency values (around 480 kHz) were caused by fracture of the fine aggregate particles. It has been observed that the number of hits and the dominant frequency of lattice-reinforced concrete were lower than those of control concrete (Liu et al. 2022).

In addition to the energy dissipated by a material during an active process, AE can be used in transmission–reception mode, where energy is generated by one sensor and received by another (resonance frequencies can exceed 1 MHz). Skibicki et al. (n.d.) examined the effect of different distributions (linear and pyramidal) and the number of layers in printed concrete, on the material's mechanical and acoustic properties. Acoustic analysis showed that ultrasonic (US) velocities and resonant frequencies are similar for printed concrete and reference concrete (similar result for US velocities in Liu et al. (2021)). No effect of the number of layers on US velocity was observed. Nevertheless, the US velocity indicated that failure of multi-layer printed samples occurs in different directions, in contrast to reference samples where cracks propagate mainly in the middle of the material. The effect of the presence of fibers (Skibicki et al. n.d.) in printed pyramidal samples was also illustrated by AE, with the highest US velocities, thus inducing high mechanical strengths (a similar result was obtained for formwork concrete by Kouta et al. (2021)).

Liu et al. (2021) conducted a study on the analysis of the anisotropy characteristics of printed concrete loaded in different directions using US propagation velocities. The results revealed a decrease in US velocities of 5.38% and 8.22% in the Y and Z directions, respectively, with loading applied in the longitudinal direction only. Results obtained by Sun et al. (2021) in the same context demonstrated that the US velocity was higher for printed samples compared to reference samples. According to the authors, this is due to a more compact microstructure and lower porosity; printed paste achieves a higher degree of final hydration, explaining the US speed result which is directly associated with high compressive strength. Ma et al. (2020) evaluated the mechanical properties of printed mortar during the initial curing phases. They obtained a result similar to the previous study, showing that printed elements reach an age of 3–7 days with a stiffness 35% higher than that of reference samples.

In conclusion, acoustic emission has been further developed for monitoring damage in hardened concrete. The application of this technique to the assessment of concrete in the fresh state has seen significant growth, highlighting the complex process of hydration. Nevertheless, the explanation of the sources of acoustic emissions is not yet fully understood. It depends on the conditions and assumptions made by the researchers during the experiments. It is therefore suggested that future research vary and record several input data (resonance frequency and sensor positions, concrete formulation and mixing process, environmental conditions). Note that in fresh concrete, in its liquid state, only longitudinal modes (where the particles in the medium vibrate parallel to the direction of propagation) propagate. Later, as the shear modulus increases (where the particles in the medium move perpendicular to the direction of propagation), shear modes and surface (Rayleigh) waves also develop. It would therefore be interesting to place AE sensors in contact with fresh concrete, so as to easily separate the different waves propagating in it. It should be noted, however, that direct contact results in strong attenuation of the waves (Iliopoulos et al. 2016), which suggests that the resonant frequency of the sensors should be chosen carefully. As far as the application of AE for 3D printing of cementitious materials is concerned, there are very few studies on this subject, spanning the period from 2020 to 2023. This research is a preliminary step toward future work to develop the basic acoustic parameters of printed structures and to deepen the analysis of crack location during loading, taking into account the anisotropy of printed concretes. It is also recommended to extend the analysis to the evaluation of mechanisms evolving during the setting of the printed material, particularly the development of microcracks at the interlayer of the material.

3.2.2. *Print control in an industrial context*

While it is relatively easy to print a single part, it is quite another challenge to maintain a certain level of quality eight hours a day, five days a week, all year round. The slightest variability in the composition of raw materials can disturb the process, sometimes leading to quality defects during printing. These disruptions can be explained by two factors: variability in the materials used and variability in the printing system and processes. Material variability can itself be explained by variations in the natural composition of raw materials, or slight variations in their manufacturing and conversion processes. It is the latter that we will be focusing on in this section. The challenge for the manufacturer is to detect these variabilities, so as to be able to compensate for their effect on final quality by adjusting formulation and/or printing parameters. This constitutes the process quality assurance plan. A quality assessment can be used to ensure that the product is within the conforming value ranges, while quality control implies a notion of eventual correction of the properties of interest in order to guarantee that their values remain within the conforming value ranges.

3D printing processes, which include a large number of sensors and actuators enabling the process to be adjusted, have a significant capacity for correcting variations in the properties of interest, compared to traditional processing methods. The question is how to modify the process parameters slightly so that the natural variabilities of the material are compensated for, thus achieving continuity of quality. Detecting these variabilities, and the modifications they entail involves a number of steps grouped together in a global industrial quality protocol.

However, the possibilities for in-process quality control differ according to the type of extrusion printing process used (see Table 3.1).

	Continuous	Discontinuous/batch
Single-component	Very poor control	Poor control
Multi-component	Medium control	High contron

Table 3.1. *Ability to perform active quality control depending on the type of system used*

The use of batch mixing means that the formula can be adjusted as the cementitious material is produced. On the other hand, the multi-component nature of the process means that the rheology of the material and its evolution can be adjusted in-line to suit the geometry to be printed. However, this increased process control implies a greater number of parameters, de facto increasing the number of control procedures to be carried out by the manufacturer. In contrast, in the case of single-component, continuously mixed printing, all control and correction procedures are the responsibility of the material manufacturer.

3.2.2.1. *Controlling ingredients on receipt*

Controlling ingredients on receipt is a vital step. In the case of printable mortars, the ingredients received by the manufacturer depend on the type of process used (see Table 3.2). Controls focus primarily on solid components: cement, secondary binders and the granular skeleton in general, since these are natural components. Additives, on the other hand, are synthetic products, and their consistency is rarely called into question.

	Continuous	Discontinuous/batch
Single-component	Premix	Premix or solid components (binders, fillers, sands, aggregates), liquid admixtures, solid admixtures
Multi-component	Premix, liquid admixtures	Premix or solid components (binders, fillers, sands, aggregates), liquid admixtures, solid admixtures

Table 3.2. *Raw material whose variability can be controlled*

Two main types of control must be considered to obtain sufficient characterization: control of binder reactivity and control of mix compactness, which drive the bulk of the material's properties in the fresh state. However, these properties are difficult to measure directly. In practice, simple rheology tests developed for the quality control of cementitious materials (mortar, grout and concrete) are carried out on reference formulas, that is, with fixed quantities of each component. The main property to be characterized is the initial shear yield stress and its temporal evolution before and after the addition of liquid admixtures in the case of multi-component technology. For this purpose, the most commonly used tests are the cone spread test (ASTM Cone) or penetration structuring tests or Vane rheometry (Nicolas et al. 2022). However, knowledge of this property alone provides only partial information on the mix's reactivity and compactness. An additional measurement would be necessary to obtain sufficient characterization (compactness and reactivity). In this case, a water demand test (Sedran 1999) can be considered, which is relatively heavy, but gives direct access to the compactness of the mix. Alternatively, measuring viscosity using a Marsh cone test (Roussel 2005), whose dependence on compactness differs from that of the shear yield stress, would make it possible to decouple the estimation of compactness and reactivity and obtain sufficient characterization for quality control of printable materials.

3.2.2.2. *Initial rheology and pumping control*

Again, depending on the type of process, the number and possibilities of rheology control differ. Initial rheology can be measured at the mixing outlet in the case of continuous mixing, or even during mixing in the case of batch mixing. Control can be carried out using the same tests as those used for suitability, or with the aid of sensors included in the printing system.

As far as pumping is concerned, there is currently no reliable universal indicator of the pumpability of a cementitious material. However, pumping is monitored using pressure transducers to track potential anomalies. Accurate in-line measurement of flow using sensors is often relatively complex to achieve. As a result, most large-scale 3D printing technologies use progressive cavity pump technology, enabling smooth dosimetric pumping. However, the efficiency of these pumps depends on the rheology of the material. It is therefore customary to calibrate the flow rate at the start of printing.

When using a variable print flow rate, it may also be important to check the linearity of the flow rate with pump rotation, which may not be achieved due to the rheo-thickening nature of mortars at high shear rates (potentially achieved at high flow rates). This calibration procedure can be carried out by simple sampling, or automatically if the system is fitted with a flow sensor.

3.2.2.3. *Control of material stiffening and printing stability*

Once pumping has started, it is necessary to adjust the consistency of the extruded material before printing begins, and then to check the quality of the extrusion and ensure that it can print a given design. To do this, it is necessary to measure the consistency of the material leaving the extruder and its structuring. Once again, this involves measuring a shear yield stress and its evolution immediately after extrusion. However, extruder exit measurements are challenging since printable materials inevitably undergo rapid changes in shear yield stress over time, making this a very active field of research.

Most often, manufacturers, especially those using single-component technologies, aim to check that the consistency of the mortar and its evolution is repeatable and greater than the minimum value expected for a given material. In the case of multi-component technologies, it is possible to adjust these values by adding the accelerator to target the structural build-up rate required for a particular design.

Measuring initial consistency

The initial consistency of the material is vital to the quality of the deposit. A yield stress that is too high will lead to a rough surface appearance and cracks, while a yield stress that is too low will cause the deposited strand to crush under its own weight. This often results in the progressive removal, layer after layer, of the robot from the previous layer, leading to the ruin of the structure. Due to the rapid evolution of this parameter at the nozzle outlet, it is not easy to measure it using the classic rheological tests discussed above. Instead, we use the slug test (Ducoulombier et al. 2021), which relies on the fact that increasing the value of the shear yield stress implies a transition from laminar flow to a drop-by-drop, slug-to-slug flow, the length of which varies proportionally with the shear yield stress. This particular property enables easy adjustment of water dosage – the main and often only parameter in continuous mixers – or of the quantity of accelerator in the event that the initial yield stress value of the initial mortar is too low.

Measuring structuring

Strictly speaking, is the structural build up affecting the evolution of all mechanical properties (in tension-shear and compression) over time? However, it is often aggregated into the evolution of the shear yield stress alone, which is relatively easy to measure. For this, the simplest method is to perform a scissometer test (hand vane test or low constant shear test) discussed in section 3.2.1.2. This test typically evaluates shear yield stresses ranging from 1 to 100 kPa (Khayat et al. 2012). Structuring is then obtained by interpolating several measurements at different rest times. This test is relatively straightforward, since it can be carried out by hand using the method presented in the previous section, after sampling the mortar at the extrusion outlet into prismatic containers. A special pocket vane geometry is also very interesting, as it enables the test to be carried out directly in situ on printed extruders (Demont et al. 2023).

Another method for assessing structure is to perform penetrometric tests, which have the advantage of being easy to automate. The various methods were presented in the previous section. However, the latter method needs to be calibrated in order to obtain a robust quantitative measurement of the shear yield stress. For certain protocols, it does, in fact, make it possible to obtain a continuous measurement of the evolution of the shear yield stress over time.

However, this measurement of aggregate structuring by shear yield stress has limitations for geometries involving complex mechanical loads, combining tensile, compressive and shear forces. Given the difficulty of fully characterizing the evolution of the strength domain over time, benchmark geometries have been developed to provide representative combinations of stresses, and to ensure that the material functions normally.

The results of these comparative studies also provide an idea of the stability obtained with a given material. This property is not directly linked to changes in the material's strength but to its elasticity. The stiffer the material, the more stable the printed part. However, the stiffness of printed materials is very difficult to measure, as it requires deformation measurements that are hard to achieve by construction manufacturers. Benchmark geometry stability is therefore a relatively simple method, providing an indirect assessment of material stiffness and its temporal evolution.

3.2.2.4. *Shrinkage control of printed samples*

The 3D concrete printing process essentially requires no formwork to guarantee geometric conformity. On the other hand, the printed material is directly exposed to variations in humidity, temperature and air flow. When the material is young, it is likely to be subject to heavy drying, particularly in dry weather. This evaporation of water from the surface of the printed material layer can have repercussions not only on cement hydration, but also on water distribution in the core of the printed material. A number of technical solutions exist to reduce the influence of these physical phenomena on the final durability of printed cementitious materials.

Three major types of shrinkage are commonly distinguished in 3D printing: endogenous shrinkage, plastic shrinkage and drying shrinkage:

– Chemical shrinkage is mainly due to the chemical activity induced by cement hydration. It occurs throughout the mass of the material, not just in contact with the surrounding environment.

– Plastic shrinkage develops in concretes that lose water while still in a plastic state, a state commonly recognized for printable cementitious materials.

– Drying shrinkage develops during hardening. The main cause is the evaporation of water contained in the porosity of the concrete from the free surfaces of the structure exposed to the ambient air.

In the case of 3D-printed concrete elements, only the final two points differ from a cast-in-place element. Drying shrinkage is conditioned by the porosity of the concrete, the humidity of the ambient air and the ratio between the exposed surface and the volume of the element. For 3D-printed structures, the multi-layer nature of the material means that the exposed surface area is considerable. What is more, the use of VMAs (viscosity modifying agents) often induces high porosity in materials during curing.

Despite the significant contribution of drying shrinkage, plastic shrinkage remains predominant, as it is mainly related to the rheological properties of the printed material. This loss of water in the fresh state is induced by surface evaporation, which can be accentuated by the difference in ambient humidity. This plastic shrinkage can also be followed by a delayed migration of water from the core of the material to areas with lower water content, those having previously undergone drying by surface evaporation. This induces a pressure gradient dominated by capillary pressure, causing water to migrate from the center of the printed material layer toward the surface exposed to air, with the aim of returning to a state of equilibrium (Keita et al. 2019). This water migration is also determined by the rheological properties of the printed materials, in particular, the permeability of the material and the viscosity of the interstitial fluid. Optimizing the compactness of the granular skeleton and adding viscosity-modifying agents helps to reduce this permeability and the associated plastic shrinkage. The addition of cellulose ethers has now become commonplace in the formulation of printable cementitious materials, as it not only achieves the rheological properties required for extrusion, but also significantly reduces surface drying (Brumaud et al. 2013).

In addition to the surface effects (cracking, crazing, dimensional variations, etc.) caused by evaporation, the lack of hydration can negatively affect the cement's hydration, thereby impacting the final strength of printed structures.

There are four main ways of curing printed materials:

– optimizing the formulation of the printable material by increasing the compactness of the granular skeleton and the viscosity of the interstitial fluid. Surface drying is possible, but is significantly reduced;

– preventing water evaporation by placing covers over the printed structures, which can also take the form of a hermetically sealed enclosure positioned on-site;

– using a curing agent, or shrinkage-reducing admixtures to reduce the surface tension of water within capillaries (Lura et al. 2007; Bentz et al. 2015; Federowicz et al. 2020);

– implementing a spraying system for the previously printed layer via a module attached to the print head (Moelich et al. 2021).

Considering the direct environment (temperature, ambient hygrometry, air flow, etc.) may eventually involve co-simulation of the additive manufacturing process during component design, so as to ensure that the required mechanical performance can be achieved. This co-simulation will consider aspects related to rheological properties in the fresh state (water migration/evaporation kinetics), properties during hydration (shrinkage and plastic deformation) and the prediction of the structure's service-life mechanical properties.

3.2.2.5. *Homogeneity control of printed materials*

Homogeneity during printing can be checked by analyzing data collected during printing. For example, droplet mass dispersions collected before printing can be used to judge material consistency during printing (Ducoulombier et al. 2021). Similarly, it is possible to use direct visual inspection or via an on-board camera to analyze defects during deposition and ensure that material extrusion remains constant during the process. The quality of the bead (geometry and surface appearance) is often a very good approximation of its homogeneity, as material heterogeneities generally result in inhomogeneous flows.

Finally, the many sensors integrated into the systems, such as pressure transducers and motor torque measurement, enable us to obtain stable curves over time during printing, highlighting the stability and homogeneity of the material's behavior during the manufacturing process.

3.2.2.6. *Printing conditions and ageing of the printed structure*

As in the precast concrete industry, controlling the aging conditions of the material is crucial. Limiting drying and controlling curing speed can be achieved in a precast-type industrial printing context to ensure optimum curing of the cementitious material. Concrete curing is a major element often used in prefabrication plants, particularly those using ultra-high-performance cementitious materials. For example, it can enhance mechanical performance and limit shrinkage and creep. The latter is more complex to apply when printing on site.

3.2.2.7. *Summary*

Due to the need to control the evolution of consistency and setting, quality protocol is naturally more important for additive manufacturing processes and is often an argument for the higher cost of this type of material. However, from our point of view, this increased control is an opportunity, particularly in the case of two-component batch printing processes. This is because the knowledge acquired about the material during preliminary trials or during the manufacturing process not only reveals any potential anomalies, but also enables the formula to be adjusted for optimum use of the material. There is even reason to believe that this could enable the use of materials whose cement content could be reduced in comparison with cementitious materials used in traditional processes, where excess cement provides the robustness required for their use in a context of low control. In the case of printable materials, a significant part of the robustness could be provided by the industrial process and quality control. Controlling curing is also an essential element in guaranteeing the quality and optimum performance of the cementitious material used.

When 3D printing in an industrial context, many parameters need to be controlled: it may be worthwhile to set up a database and analyze it to build a predictive tool for print quality and the influence of process parameters on the properties of interest. These data can also be used to build digital twins of the parts, including all the manufacturing data for the elements at each point of the printed part.

3.3. Characterization of hardened materials

3.3.1. *Specifications*

In practice, the specifications for printed materials are quite similar to those for cast cementitious materials. Depending on the mechanical loads applied to them, their mechanical properties must be characterized in one or more directions corresponding to the loading. Their durability also needs to be justified, generally by carrying out freeze–thaw cycle tests and measuring their porosity and permeability. However, further checks are often required to control the effect of printing parameters on the property of interest (mechanical, hydric and thermal). Process parameters are therefore of prime importance, and the aim is to characterize specimens whose printing parameters are representative of those used in the manufacture of the printed element.

3.3.2. *Characterization of physical properties*

When 3D printing cementitious materials, it may be necessary to determine the physical properties of the printed materials. For example, the bulk density of printed materials can be monitored to assess the homogeneity of the printed material. This involves isolating part of the material (sampling), estimating the volume and weighing the material.

Volume measurement can be complex on mortar or printed concrete beads, due to their potentially complex cross-sectional shape and lateral beading. Volume measurement by hydrostatic weighing requires the material to be sufficiently hardened, but can be carried out if the sample is protected from evaporation. The 3D scanner can be an interesting and rapid technique for accurately estimating sample volume.

Observation of the material's microstructure using fine imaging techniques such as scanning electron microscopy or micro-tomography is extremely interesting, as it can provide information on the organization of porosities within the material. However, it is very difficult to organize in an industrial or quality control context.

3.3.3. *Mechanical characterization of printed materials*

This section is based on the test protocols drawn up by the RILEM 304-ADC Technical Committee: Assessment of Additively Manufactured Concrete Materials and Structures in 2022. This was part of an international campaign in which over 20 laboratories and companies took part in the mechanical characterization of various materials printed in the cured state. These inter-laboratory and inter-company tests made it possible to evaluate different systems and materials, as well as the impact of 3D printing processes on mechanical strengths, particularly depending on the printing direction. Indeed, the superimposition of printing beads induces geometric anisotropy (the material does not necessarily have the same mechanical properties depending on the loading direction). Mechanical characterization of printed materials therefore requires specimens to be tested in different loading directions (Figure 3.4).

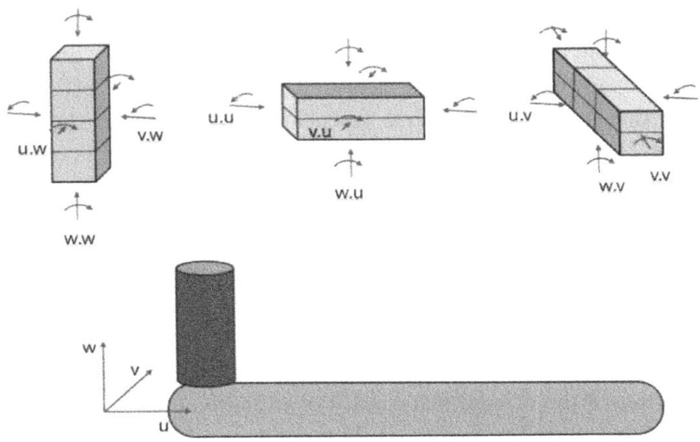

Figure 3.4. *Example of three different loading directions for prismatic flexural specimens. The loading face is shown in blue*

Due to the wide variety of printing systems used, these tests are divided into two sub-categories, one at mortar scale and the other at concrete scale. These two scales are used here to specify the size of the specimens tested. Indeed, some printing systems are capable of printing thin beads, while others are suitable for printing thicker beads. In addition, the use of 3D concrete printing varies according to requirements, whether for thin walls or volume fills. Depending on the printing system used, the printed objects will have different dimensions, as will the specimens to be mechanically tested. The remainder of this section will therefore distinguish between mortar scale and concrete scale.

The specimens detailed in the following subsections were all taken from a printed geometry. After printing, the object is protected from the outside air by being covered with a tarpaulin, then cured for the time specified for each test, generally 28 days. The day before testing, the pre-printed part is cut into several test pieces (see following sections) for mechanical testing. To ensure that the characterized material is printable, the printed object from which the test specimens are taken must have a minimum height of 25 cm for the mortar scale and 50 cm for the concrete scale.

The aim of this study is to characterize the material, not the geometry. Thus, the "beads" formed by the juxtaposed strings will be cut and removed from the test specimens (see following sections). To ensure that future test specimens can be extracted from the printed geometry, different paths are proposed depending on the width of the bead.

In this international study involving various research laboratories and companies, each participant was asked to compare printed specimens with specimens cast in place in molds, from the same material used for printing. As an option, each participant could also evaluate the impact of the cure.

3.3.3.1. *Compressive strength*

For the mortar scale, the specimens tested are 40 × 40 × 40 mm cubes, with a tolerance of ±1.0 mm. In this study, these cubes were not derived from previous bending tests (EN-196-1). The cubes are therefore extracted by cutting the printed part (Figures 3.5 and 3.6). The three possible bead orientations are tested, and each is called a series. These three series are broken down into three locations, depending on where the cube was extracted from: bottom, middle and top. As each test has to be repeated three times, there are a minimum of nine specimens per series, making a total of 27 cubic specimens.

For the concrete scale, the test specimens are 150 × 150 × 150 mm cubes, with a tolerance of 3.75 mm. All three spatial directions are also tested. As each test has to be repeated five times, this makes a total of 15 cubic test specimens for the concrete scale.

The tests in this study (Figure 3.7) are recommended to be carried out on stress-controlled presses. After each specimen has been weighed and measured, it is tested in compression to failure, at a slow, constant loading rate (2,400 N/s for mortar, 0.6 MPa/s for concrete).

Figure 3.5. *Example of specimen preparation in the Navier laboratory. Left: a portion of a printed element. Center: preparation of horizontal cut-outs. Right: preparation of vertical cut-outs*

These tests are used to determine the material's characteristic modulus of compressive strength in all three directions.

Figure 3.6. *Exarnple of cubic specimens from a printed object (Navier laboratoıy). Here, the bead lines remain visible even after cutting, depending on the mortar and system used*

Figure 3.7. *Compression test completed on cubic specimen for mortar scale*

3.3.3.2. *Flexural and tensile strength*

Three- and four-point bending tests are recommended. For the mortar scale, the specimens tested are rectangular prisms measuring 40 × 40 × 160 mm, with a tolerance of ±1.0 mm. All three possible bead orientations are tested. Each orientation test is repeated five times, for a total of 15 specimens.

For the concrete scale, the test specimens are rectangular prisms measuring 100 × 100 × 400 mm, with a tolerance of ±3.0 mm (Figure 3.8). All three possible bead orientations are tested. Each orientation test is repeated five times, for a total of 15 specimens.

The tests in this study are designed to be carried out on stress-controlled presses. After each specimen has been weighed and measured, it is tested in three-point (Figure 3.9) and four-point (Figure 3.10) bending to fracture, with a length between supports of 100 mm for the mortar scale and 300 mm for the concrete scale.

These tests are used to determine the characteristic flexural strength modulus in all three directions and to calculate a tensile strength modulus.

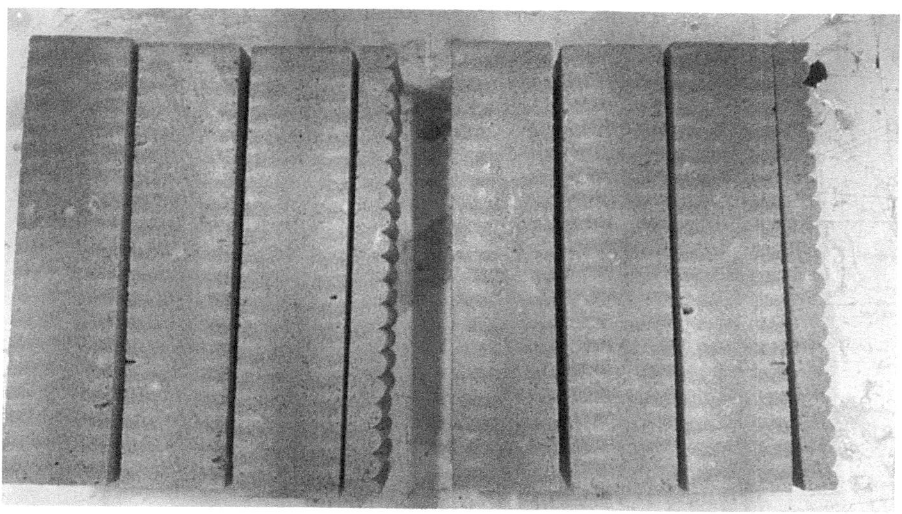

Figure 3.8. *Cutting prismatic specimens from a printed piece. Here, the traces of the bead are visible, depending on the mortar/concrete and the system used*

Figure 3.9. *Three-point bending tests at the Navier laboratory*

Figure 3.10. *Completed four-point bending tests at the Navier laboratory*

3.3.3.3. *Tensile strength and interlayer qualification*

For direct tensile tests, the specimens are of the same type and size as the prismatic specimens used for bending tests (see section 1.3.3.2).

The tests in this study are designed to be carried out on universal testing machines which can be controlled in terms of displacement. After each specimen has been weighed and measured, it is tested in tension. These tests are used to determine the characteristic tensile modulus.

To characterize the impact of the 3D printing process on a given material, the test protocol developed by RILEM has made it possible to compare results from numerous laboratories and companies around the world. The principle is to study the impact of superimposed beads in all three spatial directions, mainly by testing cubic specimens or rectangular prisms. However, this test protocol is more varied than summarized in this section.

3.4. Durability

The durability of printed concrete structures is closely linked to the properties of the materials in their fresh state and to the printing process. These parameters include, but are not limited to, construction time, extrusion potential and the rheological properties of the deposited materials.

The strong interaction between process and material is directly correlated to the final durability of the printed structures. The 3D concrete printing process is a complex additive manufacturing process which, at each stage, has a number of issues whose consequences can be observed in both the fresh and hardened states. There are three critical points of interdependence in the process itself:

– a dependency between printing speed and pumping pressure, which can have significant consequences on the porosity of the deposited material (Baz et al. 2021). However, this is less the case with two-component processes, which keep viscosity low during pumping;

– a dependency between internal and interlayer porosity with build time, or time-in-layer, that is, the time interval between printing each subsequent layer (Sanjayan et al. 2018);

– a possible but potentially major dependence of the porosity of printed concrete sections on the printing speed (Van Der Putten et al. 2018).

Matching the rheological properties of the fresh material to the entire automation chain is the main determinant of the final quality and durability of the printed concrete structure. Also, in the context of the 3D extrusion deposition printing process, there can sometimes be relatively long pauses, or a long trajectory length per layer, which means that the time elapsed between layer n and layer n+1 results in the surface drying detrimental to the adhesion of said layer n+1 (Feng et al. 2015). It is therefore important to ensure that printing parameters are chosen to limit undesirable drying. It should be noted that environmental control is also an important element in reducing surface evaporation.

Print cycle times depend on the printed geometry and the parameters of the printing process (speed, sequence, etc.). It is essential to optimize the impact of these cycle times on the manufacturing process and to take them into account in the design process for structural components.

The problem of drying at the interface, and the resulting mechanical weaknesses, is mainly due to the creation of cold joints between layers when the cycle time is too long (Panda et al. 2018). However, it has been shown that sand particle size also affects adhesion between printed layers (Zareiyan and Khoshnevis 2017). Cold joints can be temporary as the bond strength between layers increases with the creation of cement hydrates during setting. A number of studies have highlighted the value of adding an interlayer bonding agent to promote interlayer bonding. The bonding agent can be a resin (Weng et al. 2021), or even simpler, the addition of a cement paste (Marchment et al. 2019). It should also be noted that interface properties are also dependent on the type of extrusion. Infinite brick extrusion imposing zero flow at the nose outlet creates a wall effect that leads to a necessary reduction in the concentration of larger aggregates at the interface. Conversely, pressure-directed processing imposes a free-surface flow after the extruder outlet, which cancels out the wall effect imposed by the extrusion nozzle (Buswell 2020).

In any event, it will be necessary to carry out the same durability tests as for conventional cast concrete structures, while also taking into account the specific features of printed concrete (anisotropy, presence of an interlayer), the quality of which is critical for preserving the integrity of the printed structure.

Tests of resistance to chemical agents (Baz et al. 2021; Zhang et al. 2021), to chloride ion penetration (Van Der Putten et al. 2020; Malan et al. 2021; Bran-Anleu et al. 2023; Surehali et al. 2023) and to carbonation must also be carried out for printed concretes, ensuring that the printing parameters used are representative of those used in the manufacture of the structures in question.

Resistance to accidental actions such as fire or earthquakes is also a topic that requires further investigation and development (Suntharalingam et al. 2021; Delavar et al. 2023).

3.5. Conclusion

As this chapter has demonstrated, the development of 3D concrete printing is accompanied by the adaptation and creation of new tests used both in the formulation phase of a printable material to understand its material behavior as well as to implement quality control in the laboratory and throughout the value chain. However, this increased quality is an opportunity to make the best use of the material used. What is more, the 3D printing process makes it possible to aggregate a large amount of data during processing. This could in future reduce the processing uncertainty that today justifies the high safety coefficients applied during traditional processing.

Many of the advances presented in this chapter have been achieved and will serve as the basis for future standards governing the 3D concrete printing process. These will ensure that the material exhibits rheological behavior appropriate to the printing process, as well as guaranteeing the service-life behavior and durability of the printed structure.

3.6. References

Aggelis, D.G. (2011). Classification of cracking mode in concrete by acoustic emission parameters. *Mechanics Research Communications*, 38(3), 153–157.

Asprone, D., Auricchio, F., Menna, C., Mercuri, V. (2018). 3D printing of reinforced concrete elements: Technology and design approach. *Construction and Building Materials*, 165, 218–231.

Assi, L., Soltangharaei, V., Anay, R., Ziehl, P., Matta, F. (2018). Unsupervised and supervised pattern recognition of acoustic emission signals during early hydration of Portland cement paste. *Cement and Concrete Research*, 103, 216–225.

Bacharz, M., Goszczyńska, B., Trąmpczyński, W. (2015). Analysis of destructive processes in unloaded early-age concrete with the acoustic emission method. *Procedia Engineering*, 108, 245–253.

Baz, B., Aouad, G., Kleib, J., Bulteel, D., Remond, S. (2021). Durability assessment and microstructural analysis of 3D printed concrete exposed to sulfuric acid environments. *Construction and Building Materials*, 290, 123220.

Bentz, D.P., Jones, S.Z., Peltz, M.A., Stutzman, P.E. (2015). Mitigation of autogenous shrinkage in repair mortars via internal curing. *Concr. Aust*, 41(4), 35–39.

Bos, F.P., Menna, C., Pradena, M., Kreiger, E., da Silva, W.R.L., Rehman, A.U., Weger, D., Wolfs, R.J.M., Zhang, Y., Ferrara, L. et al. (2022). The realities of additively manufactured concrete structures in practice. *Cement and Concrete Research*, 156, 106746.

Bran-Anleu, P., Wangler, T., Nerella, V.N., Mechtcherine, V., Trtik, P., Flatt, R.J. (2023). Using micro-XRF to characterize chloride ingress through cold joints in 3D printed concrete. *Materials and Structures*, 56(3), 51.

Brumaud, C., Bessaies-Bey, H., Mohler, C., Baumann, R., Schmitz, M., Radler, M., Roussel, N. (2013). Cellulose ethers and water retention. *Cement and Concrete Research*, 53, 176–184.

Buswell, R.A., da Silva, W.R.L., Bos, F.P., Schipper, H.R., Lowke, D., Hack, N. (2020). A process classification framework for defining and describing digital fabrication with concrete. *Cement and Concrete Research*, 134, 106068.

Chotard, T.J., Barthelemy, J., Smith, A., Gimet-Breart, N., Huger, M., Fargeot, D., Gault, C. (2001). Acoustic emission monitoring of calcium aluminate cement setting at the early age. *Journal of Materials Science Letters*, 20, 667–669.

Delavar, M.A., Chen, H., Sideris, P. (2023). Seismic design methodology for 3D printed concrete buildings. *Cityscape*, 25(1), 177–197.

Demont, L., Mesnil, R., Ducoulombier, N., Caron, J.-F. (2023). Affordable inline structuration measurements of printable mortar with a pocket shear vane. *Construction and Building Materials*, 408, 133602.

Drouillard, T.F. (1994). Acoustic emission: The first half century. Report, EG & G Rocky Flats, Inc., Golden, Colorado.

Ducoulombier, N., Mesnil, R., Carneau, P., Demont, L., Bessaies-Bey, H., Caron, J.-F., Roussel, N. (2021). The "Slugs-test" for extrusion-based additive manufacturing: Protocol, analysis and practical limits. *Cement and Concrete Composites*, 121, 104074.

Dzaye, E.D., De Schutter, G., Aggelis, D.G. (2018). Study on mechanical acoustic emission sources in fresh concrete. *Archives of Civil and Mechanical Engineering*, 18, 742–754.

Dzaye, E.D., De Schutter, G., Aggelis, D.G. (2020). Monitoring early-age acoustic emission of cement paste and fly ash paste. *Cement and Concrete Research*, 129, 105964.

Estellé, P., Lanos, C., Perrot, A., Amziane, S. (2008a). Processing the vane shear flow data from Couette analogy. *Applied Rheology*, 18(3), 34037-1–34037-6.

Estellé, P., Lanos, C., Perrot, A. (2008b). Processing the Couette viscometry data using a Bingham approximation in shear rate calculation. *Journal of Non-Newtonian Fluid Mechanics*, 154(1), 31–38.

Federowicz, K., Kaszyńska, M., Zieliński, A., Hoffmann, M. (2020). Effect of curing methods on shrinkage development in 3D-printed concrete. *Materials*, 13(11), 2590.

Feng, P., Meng, X., Chen, J.-F., Ye, L. (2015). Mechanical properties of structures 3D printed with cementitious powders. *Construction and Building Materials*, 93, 486–497.

Feys, D., Wallevik, J.E., Yahia, A., Khayat, K.H., Wallevik, O.H. (2013). Extension of the Reiner–Riwlin equation to determine modified Bingham parameters measured in coaxial cylinders rheometers. *Materials and Structures*, 46, 289–311.

Feys, D., Cepuritis, R., Jacobsen, S., Lesage, K., Secrieru, E., Yahia, A. (2018). Measuring rheological properties of cement pastes: Most common techniques, procedures and challenges. *RILEM Technical Letters*, 2, 129–135.

Feys, D., Sonebi, M., Amziane, S., Djelal, C., El-Cheikh, K., Fataei, S., Greim, M., Ivanova, I., Keller, H., Khayat, K. (2023). RILEM TC 266-MRP: Round-robin rheological tests on high performance mortar and concrete with adapted rheology – Rheometers, mixtures and procedures. *Materials and Structures*, 56(4), 90.

Fournier, C. (2018). Panorama des normes et documents normatifs dans le domaine des essais non destructifs. AFNOR Report, Paris.

Iliopoulos, S.N., Khattabi, Y.E., Aggelis, D.G. (2016). Towards the establishment of a continuous nondestructive monitoring technique for fresh concrete. *Journal of Nondestructive Evaluation*, 35, 1–11.

Ivanova, I. and Mechtcherine, V. (2020). Possibilities and challenges of constant shear rate test for evaluation of structural build-up rate of cementitious materials. *Cement and Concrete Research*, 130, 105974.

Jacquet, Y., Picandet, V., Rangeard, D., Perrot, A. (2020). Gravity induced flow to characterize rheological properties of printable cement-based materials. *RILEM Technical Letters*, 5, 150–156.

Jacquet, Y., Perrot, A., Picandet, V. (2021a). Assessment of asymmetrical rheological behavior of cementitious material for 3D printing application. *Cement and Concrete Research*, 140, 106305.

Jacquet, Y., Picandet, V., Perrot, A. (2021b). Characterization of tensile behavior of fresh cementitious materials. *ACI Materials Journal*, 118(6), 217–226.

Jacquet, Y., Perrot, A., Picandet, V. (2022). Monitoring strain using digital image correlation during compressive and tensile loading: Assessment of critical strain of cement-based materials containing VMA. In *RILEM International Conference on Concrete and Digital Fabrication*. Springer, Heidelberg.

Keita, E., Bessaies-Bey, H., Zuo, W., Belin, P., Roussel, N. (2019). Weak bond strength between successive layers in extrusion-based additive manufacturing: Measurement and physical origin. *Cement and Concrete Research*, 123, 105787.

Khayat, K.H., Omran, A.F., Naji, S., Billberg, P., Yahia, A. (2012). Field-oriented test methods to evaluate structural build-up at rest of flowable mortar and concrete. *Materials and Structures*, 45, 1547–1564.

Kouta, N., Saliba, J., Saiyouri, N. (2021). Fracture behavior of flax fibers reinforced earth concrete. *Engineering Fracture Mechanics*, 241, 107378.

Liu, C., Yue, S., Zhou, C., Sun, H., Deng, S., Gao, F., Tan, Y. (2021). Anisotropic mechanical properties of extrusion-based 3D printed layered concrete. *Journal of Materials Science*, 56(30), 16851–16864.

Liu, J., Kanwal, H., Tang, C., Hao, W. (2022). Study on flexural properties of 3D printed lattice-reinforced concrete structures using acoustic emission and digital image correlation. *Construction and Building Materials*, 333, 127418.

Lootens, D., Jousset, P., Martinie, L. (2009). Yield stress during setting of cement pastes from penetration tests. *Cem. Concr. Res.*, 39. doi: 10.1016/j.cemconres.2009.01.012.

Lura, P., Pease, B., Mazzotta, G.B., Rajabipour, F., Weiss, J. (2007). Influence of shrinkage-reducing admixtures on development of plastic shrinkage cracks. *ACI Materials Journal*, 104(2), 187.

Lura, P., Couch, J., Jensen, O.M., Weiss, J. (2009). Early-age acoustic emission measurements in hydrating cement paste: Evidence for cavitation during solidification due to self-desiccation. *Cement and Concrete Research*, 39(10), 861–867.

Ma, G., Li, Y., Wang, L., Zhang, J., Li, Z. (2020). Real-time quantification of fresh and hardened mechanical property for 3D printing material by intellectualization with piezoelectric transducers. *Construction and Building Materials*, 241, 117982.

Malan, J.D., van Rooyen, A.S., van Zijl, G.P. (2021). Chloride induced corrosion and carbonation in 3D printed concrete. *Infrastructures*, 7(1), 1.

Marchment, T., Sanjayan, J., Xia, M. (2019). Method of enhancing interlayer bond strength in construction scale 3D printing with mortar by effective bond area amplification. *Materials & Design*, 169, 107684.

Mazhoud, B., Perrot, A., Picandet, V., Rangeard, D., Courteille, E. (2019). Underwater 3D printing of cement-based mortar. *Construction and Building Materials*, 214, 458–467.

McLaskey, G.C., Glaser, S.D., Grosse, C.U. (2007). Integrating broadband high-fidelity acoustic emission sensors and array processing to study drying shrinkage cracking in concrete. In *Sensors and Smart Structures Technologies for Civil, Mechanical, and Aerospace Systems 2007*. SPIE, Colorado.

Mechtcherine, V., Nerella, V.N., Kasten, K. (2014). Testing pumpability of concrete using sliding pipe rheometer. *Construction and Building Materials*, 53, 312–323.

Mechtcherine, V., Buswell, R., Kloft, H., Bos, F.P., Hack, N., Wolfs, R., Sanjayan, J., Nematollahi, B., Ivaniuk, E., Neef, T. (2021). Integrating reinforcement in digital fabrication with concrete: A review and classification framework. *Cement and Concrete Composites*, 119, 103964.

Mechtcherine, V., Van Tittelboom, K., Kazemian, A., Kreiger, E., Nematollahi, B., Nerella, V.N., Santhanam, M., De Schutter, G., Van Zijl, G., Lowke, D. (2022). A roadmap for quality control of hardening and hardened printed concrete. *Cement and Concrete Research*, 157, 106800.

Mix, P.E. (2005). *Introduction to Nondestructive Testing: A Training Guide*. John Wiley & Sons, New York.

Moelich, G.M., Kruger, J., Combrinck, R. (2021). Modelling the interlayer bond strength of 3D printed concrete with surface moisture. *Cement and Concrete Research*, 150, 106559.

Nerella, V.N. and Mechtcherine, V. (2018). Virtual sliding pipe rheometer for estimating pumpability of concrete. *Construction and Building Materials*, 170, 366–377.

Nicolas, R., Richard, B., Nicolas, D., Irina, I., Temitope, K.J., Dirk, L., Viktor, M., Romain, M., Arnaud, P., Ursula, P. (2022). Assessing the fresh properties of printable cement-based materials: High potential tests for quality control. *Cement and Concrete Research*, 158, 106836.

Ohno, K. and Ohtsu, M. (2010). Crack classification in concrete based on acoustic emission. *Construction and Building Materials*, 24(12), 2339–2346.

Ohtsu, M. (2006). Quantitative AE techniques standardized for concrete structures. *Advanced Materials Research*, 13, 183–192.

Ohtsu, M. and Grosse, C. (2008). *Acoustic Emission Testing: Basics for Research-Applications in Civil Engineering*. Springer, Heidelberg.

Panda, B., Paul, S.C., Mohamed, N.A.N., Tay, Y.W.D., Tan, M.J. (2018). Measurement of tensile bond strength of 3D printed geopolymer mortar. *Measurement*, 113, 108–116.

Perrot, A. and Amziane, S. (2024). Concrete rheometers. In *Measuring Rheological Properties of Cement-based Materials: State-of-the-Art Report of the RILEM Technical Committee 266-MRP*, Sonebi, M. and Feys, D. (eds). Springer Nature, Cham. doi: 10.1007/978-3-031-36743-4_3.

Perrot, A., Mélinge, Y., Rangeard, D., Micaelli, F., Estellé, P., Lanos, C. (2012). Use of ram extruder as a combined rheo-tribometer to study the behaviour of high yield stress fluids at low strain rate. *Rheologica Acta*, 51, 743–754.

Perrot, A., Rangeard, D., Pierre, A. (2016). Structural built-up of cement-based materials used for 3D-printing extrusion techniques. *Materials and Structures*, 49, 1213–1220.

Pott, U. and Stephan, D. (2021). Penetration test as a fast method to determine yield stress and structural build-up for 3D printing of cementitious materials. *Cement and Concrete Composites*, 121, 104066.

Reiter, L., Wangler, T., Roussel, N., Flatt, R.J. (2022). Slow penetration for characterizing concrete for digital fabrication. *Cement and Concrete Research*, 157, 106802.

RILEM Technical Committee (Masayasu Ohtsu) (2010). Recommendation of RILEM TC 212-ACD: Acoustic emission and related NDE techniques for crack detection and damage evaluation in concrete: Test method for damage qualification of reinforced concrete beams by acoustic emission. *Materials and Structures*, 43(9), 1183–1186.

Roussel, N. (2005). Steady and transient flow behaviour of fresh cement pastes. *Cement and Concrete Research*, 35(9), 1656–1664.

Roussel, N. (2018). Rheological requirements for printable concretes. *Cement and Concrete Research*, 112, 76–85.

Roussel, N. and Le Roy, R. (2005). The Marsh cone: A test or a rheological apparatus? *Cement and Concrete Research*, 35, 823–830. doi: 10.1016/j.cemconres.2004.08.019.

Sanjayan, J.G., Nematollahi, B., Xia, M., Marchment, T. (2018). Effect of surface moisture on inter-layer strength of 3D printed concrete. *Construction and Building Materials*, 172, 468–475.

Secrieru, E., Khodor, J., Schröfl, C., Mechtcherine, V. (2018). Formation of lubricating layer and flow type during pumping of cement-based materials. *Construction and Building Materials*, 178, 507–517.

Secrieru, E., Mohamed, W., Fataei, S., Mechtcherine, V. (2020). Assessment and prediction of concrete flow and pumping pressure in pipeline. *Cement and Concrete Composites*, 107, 103495.

Sedran, T. (1999). *Rhéologie et rhéométrie des betons. Application aux bétons autonivelants*. Ecole Nationale des Ponts et Chaussées, Paris.

Shen, G., Zhang, J., Lackner, G. (2020). International acoustic emission standard analysis and development outlook. *Insight-Non-Destructive Testing and Condition Monitoring*, 62(12), 724–734.

Shiotani, T., Bisschop, J., Van Mier, J.G.M. (2003). Temporal and spatial development of drying shrinkage cracking in cement-based materials. *Engineering Fracture Mechanics*, 70(12), 1509–1525.

Skal's'kyi, V.R., Koval, P.M., Serhienko, O.M., Lotots'kyi, Y.L. (2004). Investigation of the solidification of concrete according to the signals of acoustic emission. *Materials Science*, 40, 698–701.

Skibicki, S., Dvořák, R., Pazdera, L., Topolář, L., Kocáb, D., Alexa, M., Cendrowski, K., Hoffmann, M. (n.d.). Anisotropic mechanical properties of 3D printed mortar by standard flexural and compression test and acoustic emission. *SSRN*, 4543414.

Sun, J., Huang, Y., Aslani, F., Wang, X., Ma, G. (2021). Mechanical enhancement for EMW-absorbing cementitious material using 3D concrete printing. *Journal of Building Engineering*, 41, 102763.

Suntharalingam, T., Upasiri, I., Gatheeshgar, P., Poologanathan, K., Nagaratnam, B., Rajanayagam, H., Navaratnam, S. (2021). Fire resistance of 3D printed concrete composite wall panels exposed to various fire scenarios. *Journal of Structural Fire Engineering*, 12(3), 377–409.

Surehali, S., Tripathi, A., Nimbalkar, A.S., Neithalath, N. (2023). Anisotropic chloride transport in 3D printed concrete and its dependence on layer height and interface types. *Additive Manufacturing*, 62, 103405.

Tavangar, T., Hosseinpoor, M., Yahia, A., Khayat, K.H. (2022). Novel tri-viscous model to simulate pumping of flowable concrete through characterization of lubrication layer and plug zones. *Cement and Concrete Composites*, 126, 104370.

Thirumalaiselvi, A. and Sasmal, S. (2019). Acoustic emission monitoring and classification of signals in cement composites during early-age hydration. *Construction and Building Materials*, 196, 411–427.

Van Den Abeele, K., Desadeleer, W., De Schutter, G., Wevers, M. (2009). Active and passive monitoring of the early hydration process in concrete using linear and nonlinear acoustics. *Cement and Concrete Research*, 39(5), 426–432.

Van Der Putten, J., De Schutter, G., Van Tittelboom, K. (2018). The effect of print parameters on the (micro) structure of 3D printed cementitious materials. In *RILEM International Conference on Concrete and Digital Fabrication*. Springer, Cham.

Van Der Putten, J., De Volder, M., Van den Heede, P., De Schutter, G., Van Tittelboom, K. (2020). 3D printing of concrete: The influence on chloride penetration. In *RILEM International Conference on Concrete and Digital Fabrication*. Springer, Cham.

Varela, H., Barluenga, G., Perrot, A. (2023). Extrusion and structural build-up of 3D printing cement pastes with fly ash, nanoclays and VMAs. *Cement and Concrete Composites*, 142, 105217.

Wallevik, O.H., Feys, D., Wallevik, J.E., Khayat, K.H. (2015). Avoiding inaccurate interpretations of rheological measurements for cement-based materials. *Cement and Concrete Research*, 78, 100–109.

Weng, Y., Li, M., Wong, T.N., Tan, M.J. (2021). Synchronized concrete and bonding agent deposition system for interlayer bond strength enhancement in 3D concrete printing. *Automation in Construction*, 123, 103546.

Wolfs, R.J.M., Bos, F.P., Salet, T.A.M. (2018). Early age mechanical behaviour of 3D printed concrete: Numerical modelling and experimental testing. *Cement and Concrete Research*, 106, 103–116.

Yahia, A. and Perrot, A. (2024). Measuring procedures. In *Measuring Rheological Properties of Cement-based Materials: State-of-the-Art Report of the RILEM Technical Committee 266-MRP*, Sonebi, M. and Feys, D. (eds). Springer, Cham. doi: 10.1007/978-3-031-36743-4_4.

Yahia, A., Perrot, A., Feys, D., Khayat, K., Sonebi, M., Kawashima, S., Schmidt, W. (2023). Viscoelastic properties of fresh cement paste: Measuring procedures and influencing parameters. *Rilem Technical Letters*, 8, 23–30.

Yue, J.G., Beskos, D.E., Feng, C., Wu, K. (2022). Hardened fracture characteristics of printed concrete using acoustic emission monitoring technique. *Construction and Building Materials*, 361, 129684.

Zareiyan, B. and Khoshnevis, B. (2017). Interlayer adhesion and strength of structures in contour crafting-effects of aggregate size, extrusion rate, and layer thickness. *Automation in Construction*, 81, 112–121.

Zhang, Y., Zhang, Y., Yang, L., Liu, G., Chen, Y., Yu, S., Du, H. (2021). Hardened properties and durability of large-scale 3D printed cement-based materials. *Materials and Structures*, 54(1), 45.

4

Alternative Printing Methods for Cementitious Materials

Alexandre PIERRE[1] and Arnaud PERROT[2]

[1] *CY Cergy Paris Université, France*

[2] *IRDL, Université de Bretagne Sud, Lorient, France*

4.1. Introduction

Robotic concrete construction goes beyond the technique of depositing successive layers or filaments, which then solidify. In the field of construction materials, 3D printing technologies are multiple and not limited to extrusion 3D printing. Spatial resolution and absolute freedom of form can be achieved by alternative 3D printing techniques, which we will discuss in this chapter.

Large-scale 3D printing projects for new housing, structures (like pedestrian bridges) or smaller scale equipment and street furniture are currently being carried out using 3D extrusion printing. Their number is currently growing, and they require R&D before they can be marketed. In terms of the number of construction sites, the main applications of 3D printing are limited to relatively simple shapes with little variation in cross-section, and optimized material quantities that are not always financially or environmentally profitable. Indeed, a relatively thin filament thickness is required to achieve an environmental gain (Kuzmenko et al. 2022). The cost of customized formwork, stable productivity since the 1950s and the unusual use of robotics also limit a massive breakthrough of this process on the construction

For a color version of all figures in this chapter, see www.iste.co.uk/perrot/3dconcrete.zip.

3D Concrete Printing,
coordinated by Arnaud PERROT and Yohan JACQUET. © ISTE Ltd 2025.

market. As early as 2016, some authors were already raising questions about the mass use of 3D printing in civil engineering (Wu et al. 2016).

Bearing in mind that concrete construction elements are legally subject to compliance with Eurocodes standards, the freedom of design and complexity of form brought about by digitalization are reduced. Printed concrete elements are generally used as lost formwork, or as temporary (or even permanent) supports made from other materials. This opens up a wide range of possibilities, including the alternative techniques described in this chapter.

The desire to create elements assembled in the form of structures from materials in their fresh state, driven by technical and environmental choices, has prompted some players in the field to extend their technologies beyond the traditional extrusion triptych: liquid shaping, deposition and solidification. The main difficulty not fully solved by extrusion 3D printing is to form the material without spatial constraints and to add structural reinforcements. There is a real need for innovation to meet the challenge of innovative, disruptive 3D printing in the construction sector. Alternative techniques for additive manufacturing of construction materials need to be explored.

In this chapter, a number of alternative additive manufacturing processes will be presented. The impact of using alternative 3D printing methods could be immense: building components with customized properties and optimized shapes can reduce waste and improve the energy efficiency of buildings. The opportunity to actually have materialization in space with these techniques could be interesting for the construction sector.

The freedom of form of a printed work or element, both for practical, functional and aesthetic reasons, is the main ambition of emerging 3D printing techniques. What are the current possibilities offered by alternative digital manufacturing techniques to break free from a process dominated solely by the superimposition of printed filaments? Do alternative techniques meet the main objectives of 3D printing, which are currently unattainable with the extrusion technique? These include reducing the amount of material used, minimizing waste and reusing materials, designing elements for circularity and reuse, manufacturing functional geometries through their freedom of form and increasing automation in prefabrication.

Other possibilities are yet to be explored, both for housing renovation and refurbishment, and for urban design and structural ornamentation. In this chapter, the alternative 3D printing processes to extrusion will be explored, presenting the most well-documented and recent alternative techniques. The main aim of this summary chapter is to highlight the current strengths and limitations of alternative additive manufacturing techniques applied to cementitious materials.

4.2. Methods with supports

Printing concrete material alone is sometimes not enough to meet the functional (structural, thermal or aesthetic) requirements of the printed element. One way of ensuring that the material is held in place when it is placed by the robot is to use a support. This support initially acts as a holding device for the material, leaving the printer head. It can be solid, non-deformable or flexible, liquid, moving or fixed. It is also possible to use a so-called sacrificial support, which can be reused. Sacrificial supports are commonly used in polymer 3D printing, but remain confidential in the field of civil engineering. We can classify techniques using solid supports or sacrificial supports, in situ.

On the one hand, the advantage of a solid support is that it can be re-used, but this contradicts the advantage of 3D printing's freedom of form, particularly in the original sense of formwork, which is to repeat modules or elements in series. On the other hand, printed formwork that remains in place, or in situ formwork, is ideally suited to bringing functionality to the material to be poured into it. It must go beyond the mere aesthetic or geometric aspect that a conventional extrusion process can satisfy. It can provide structural support (reinforcement or textile fibers), acoustic and thermal comfort (foam), help against carbonation (polymer) or can improve durability on the surface exposed to external environmental stresses.

4.2.1. *In situ printed supports (permanent formwork)*

4.2.1.1. *Semi-permeable metal support*

Alternative 3D printing techniques aim to create more complex shapes than extrusion. However, the infinite possibilities of printed structures are still a long way off. One reason is that the material leaving the nozzle has relatively low mechanical strength compared with other printable materials such as metal alloys.

From an architectural point of view, the manual placement of metal reinforcement as sacrificial formwork was, for example, a solution used for the Taichung Metropolitan Opera House (Figure 4.1). The introduction of robotized reinforcement placement has modernized the manufacture of this type of structure, with permeable formwork and the introduction of curved, slender geometries. Historically, this process, which ultimately involves printing a mesh or lattice permeable formwork, known as "mesh mould", was first developed using polymers from 2013 onward (Boyd IV et al. 2020; Jipa and Dillenburger 2022). Figure 4.1 shows the device developed at ETH Zurich for printing the formwork into which the concrete will be poured (Hack and Lauer 2014; Jipa and Dillenburger 2022). It consists of a printhead capable of bending (front part), cutting and welding steel reinforcement bars (rear part).

The scientific challenges of this technique are, on one scale, the same as those developed several years ago by flowing concrete through reinforcement cages (Vasilic et al. 2011, 2016). The present case is more complex, because instead of allowing concrete to pass through the reinforcement, the process requires the material to be partially blocked, then equivalent to a "liquid stone". In fact, it is necessary to fill the space inside the formwork with reinforcement and cover it on the visible side of the element in order to preserve its service life. The characteristic size of the mesh and its permeability must match the characteristic size of the material in place. This type of element is suitable when external conditions are not extreme or do not have a high impact on the durability of the structure, due to the finer coverage of the reinforcement, rather than with traditional concrete.

Figure 4.1. *Left: interior and exterior architecture of the Taichung Metropolitan Opera House (© Fcuk1203 © 李家宇); right: manual and robotic rebar installation (© Gramazio Kohler Research)*

4.2.1.2. *Semi-permeable fiber support*

An adaptation of the mesh mold technique was developed in 2018 using a fibrous mortar sprayed onto a permeable carbon fiber structure (Taha et al. 2019). This technique, known as “aerocrete”, enables curved and very fine shapes. The right compromise between the permeability of the formwork and the size of the fibers used is necessary to obtain adequate encapsulation of the structure. However, the quantities of material lost during spraying and the poor surface quality have, for the time being, kept this technique at the prototype stage (Szabo 2020). The disadvantages of this technique include the need to adapt the spraying process to the size and geometry of the mesh (speed, quantity, trajectory, flow rate, etc.). The curing time of the sprayed concrete is also an important parameter for potential industrialization of the process.

The “Knitcrete” process, developed in 2016, uses a 3D printer to produce a textile structure using a technique similar to robotic knitting. The textile fibers are then assembled and tensioned. One side of the textile is sprayed with a quick-setting calcium aluminate cement, and concrete is then added to the other side (Popescu et al. 2018; Popescu 2019). The advantages of these structures are lightness and easy transport of the textile elements, while avoiding complex assemblies. Concrete curing times also need to be adapted for on-site or outdoor applications. The mechanical contribution of the textile structure remains to be assessed.

Figure 4.2 shows elements obtained with these two techniques.

Figure 4.2. *Top left: model of robot trajectory; bottom left: textile structure sewn and fully covered; right: robotic spraying of building material onto textile structure (photographs from Rennen et al. 2023)*

4.2.1.3. *Foam support*

Future work in 3D printing will increasingly incorporate the improvement of properties on the scale of building physics. To this end, mineral foams are cellular materials generally used for thermal insulation, thanks to their air cavities (the material's main constituent). They are mainly used to limit heat loss in buildings. However, traditional methods require considerable manpower to apply this type of insulation on site. An alternative 3D printing technique is to use these binders as permanent or temporary formwork. Figure 4.3 illustrates a process involving the 3D printing of mineral foam modules, which are left in place to provide acoustic or thermal functionalities. It is also possible to remove these modules to make room for the structure's cavities and lighten the structure.

The state of the art by Bedarf et al. (2021) reviews the applications and processes used and highlights two advantages. The first is the lightness of the structure due to the air cavities and the second is environmental, with improved thermal performance and reduced use of materials. There are still a number of unresolved technological issues to be addressed, including material formulation, foaming control, printing techniques and on-site developments.

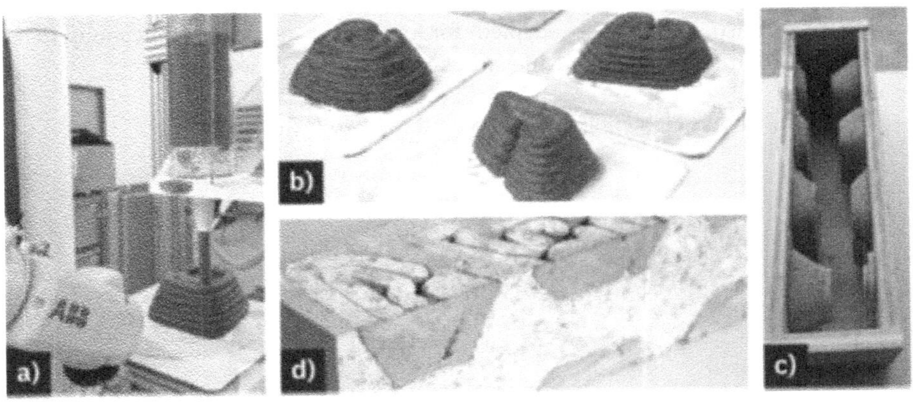

Figure 4.3. *Self-supporting mineral foam formwork system. a) 3D printing of mineral foam components; b) resulting mineral foam components; c) mineral foam elements attached to wooden formwork; d) mineral foam elements embedded in concrete (photos © Bedarf, P., Calvo-Barentin, C., Schulte, D.M., from Bedarf et al. (2023), licensed under CC BY 4.0)*

The main challenge is to stabilize the material's properties and control the evolution of its porosity over time and during the process. When a foam is set in motion during processing, it undergoes forces, notably shear forces (Goyon et al.

2010). It must also resist compressive forces during installation, while remaining compressible. This antagonism requires a thorough understanding of the material's behavior and properties. The nature of the binder, the size, morphology and distribution of the cavities are parameters to be controlled when using this technique (Bedarf et al. 2021). To this end, the porous structure is generally obtained by combining mechanical and chemical actions. During mixing, air is incorporated and the expansion of the mixture is linked to the amount of foaming agents added to the mix. As a reminder, the order of magnitude of the thermal conductivity of "traditional" concrete is 0.5 $W \cdot m^{-1} \cdot K^{-1}$ for a density of 250 kg/m^3, whereas "foamed" concrete can reach a density close to 250 kg/m^3 with a thermal conductivity of 0.065 $W \cdot m^{-1} \cdot K^{-1}$.

4.2.2. *Mobile support*

3D extrusion printing sometimes generates defects at the interface of the printed layers, known as cold joints (Perrot et al. 2021). This difficulty in assembling and reuniting layers means that the structure will have a limited lifespan, depending on external stresses. What is more, in the event of late layer overlap, drying phenomena will create micro-structural defects, which can then degenerate the primary functionality of the printed element.

A different approach, proposed back in 2012 by ETH Zurich and known as Smart Dynamic Casting, involves shaping concrete with a (sliding) formwork whose (dynamic) trajectory is robotized. The concrete enters the formwork in a liquid state and emerges from the formwork in a solid state in just a few minutes. The advantage of this technique lies in its quasi-continuous process. Compared with conventional extrusion, there are no defects at the interface. The material must be structured, hydrated and set to suit the trajectory of the dynamic formwork. As this trajectory is generally vertical, the correlation of the two speeds – material structuring and robot movement – is essential.

Technologically, in the early work developed at ETH Zurich and explained in (Lloret-Fritschi et al. 2017, 2020), a flexible, controllable membrane was arranged at the mold outlet. This made it possible to obtain sections scaled to one-fifth the diameter of the mold. The main challenge during the process is to reduce friction when lifting the mold. In fact, this work has shown that the weight of the formwork decreases sharply during vertical movement, synonymous with the phenomenon of friction (Szabo et al. 2019b, 2020b). Lubricating systems at the interface are therefore essential for the success of these structures.

Compared with conventional slipforming, the SDC technique opens up two possibilities: transforming the geometry and volume of the final form, thanks to the process (robot rotation, adjustable speed, etc.). The speed of the formwork must be adapted to the hydration rate of the concrete, so that it can support its own weight and that inside the formwork. Yield stress values for material flow can therefore be higher than those used in extrusion. The success of this process depends on the compromise between the force due to gravity and the frictional force generated along the mold during movement. Buckling of the structure is also a difficulty to be taken into account when designing the shape.

A limitation to the use of an articulated arm, synonymous with freedom of form in the space allocated to the arm's movement, is sometimes not possible because the weight of the formwork and material cannot be supported by the robot. This calls for the use of gantry systems, which reduce architectural freedom. The systems developed offer two options: either modify the cross-section of the material by adapting the movement or by placing devices at the nozzle outlet, or modify the formwork, automatically inducing a change in the shape of the material. The second, simpler option reduces stresses on the material, but can lead to significant frictional effects. Figure 4.4 illustrates the evolution of the dynamic slipform technique according to Lloret-Fritschi et al. (2020).

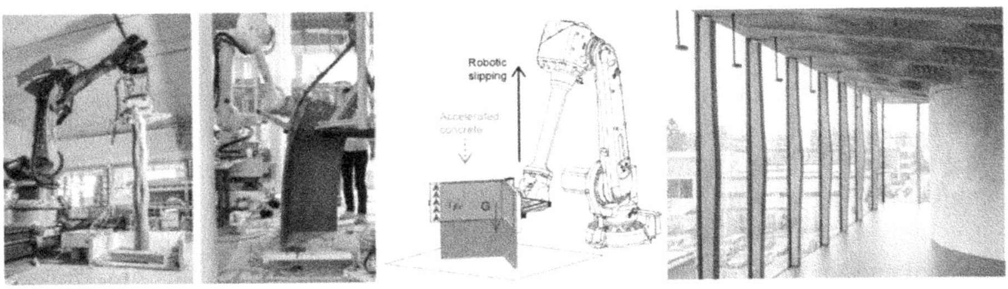

Figure 4.4. *Examples of the SDC technique. From left to right: slipforming with changes in cross-section during elevation of the printed column; thin structures, schematic representation of force balance during the process (photos from Szabo et al. 2020b, licensed under CC BY 4.0); beams produced using the SDC technique (photo from Lowke et al. 2024, licensed under CC BY 4.0)*

4.2.3. *Sacrificial support*

4.2.3.1. *Flexible tension support*

Flexible supports of various materials can also be filled with concrete deposited along a robotized trajectory. This process, also developed in Zurich as a continuation of the work on smart dynamic casting, is slightly different. The formwork, made of textile, paper or plastic, is tensioned but remains static, while the concrete is poured following the robot's trajectory. A support structure for these flexible molds is installed beforehand to give the object its final shape. Developed in Szabo's thesis work (Szabo 2020), the concrete must have an "over-demand" structure and set to reduce the pressure exerted on the formwork, which is too thin and of low rigidity to support the weight of the material. Figure 4.5 shows, for example, a flexible polyethylene (PE) support tensioned between two wooden supports. Filling speeds can be rapid of the order of 2.5 m/h (Lloret-Fritschi et al. 2020).

We also note that the formwork device here is affordable and easy to set up. The difficulties lie in rheological and chemical control of the material poured into the flexible formwork, and in appropriate robot control. It is essential to formulate a rapidly hydrating material due to the low capacity of flexible supports to withstand hydrostatic pressure. To achieve this, aluminum sulfate must be added to the formulation. It promotes rapid ettringite formation and increases the material's yield stress. We can see from Figure 4.5 that surface aspects are obviously correlated with the material used (smooth surface for PE, opaque for cardboard, slightly rough for geotextile).

Figure 4.5. *Robotic concrete casting in PE film under tension (photographs from Szabo et al. 2019a, licensed under CC BY 4.0)*

4.2.3.2. *Thin polymer support (eggshell)*

The recycling of plastic materials makes it possible to use extruded polymer as formwork. This temporary formwork until the concrete solidifies is then more easily printable with high resolution of the final part, thanks to hindsight on 3D printing of polymer-based material. This process is referred to as "eggshell". The principle is simple: print the thinnest possible skin or layer at high resolution and place the concrete on it, either upstream or simultaneously with the concrete pour (Lloret-Fritschi et al. 2020). The challenge is to adapt the mold filling speed to the mechanical performance of the polymer formwork and the structural build-up of the concrete, due to its so-called thixotropy, in order to limit hydrostatic pressure. The process does not allow rapid production due to the relatively long time required to print the polymer. Similarly to each alternative technique with a solid support material, the placement of structural reinforcements is possible. Figure 4.6 shows several examples of parts produced using this technique, including a reinforced column 1 m 80 cm high.

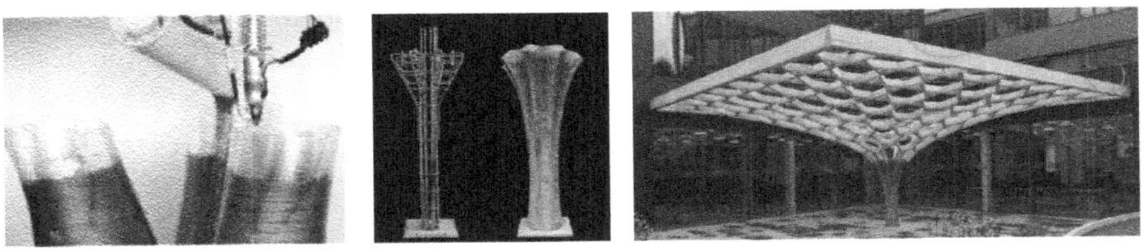

Figure 4.6. *Illustration of projects using the polymer printing technique as formwork; left: simultaneous printing of polymer and pouring of cementitious material; middle and right: printed column with reinforcement and use as part of the "future tree" project; photos taken from Jipa and Dillenburger (2022) and Lowke et al. (2024)*

It is also possible to use substrates with a thickness of less than 1 mm, but a counterpressure system is required to ensure stability when the concrete is placed. In this case, the concrete used is not formulated to harden and hydrate very quickly, as the pressure is taken up by a sand bank placed around the printed mold. An example is shown in Figure 4.7. Here, several dozen elements were 3D printed with a biopolymer to make formworks (Jipa et al. 2019). These elements were then assembled by cold welding. Then, concrete, fluid enough not to clog the channels of the printed mold shapes, was injected. The sand, rising parallel to the level of the concrete, was placed within the structure, to relieve the stresses imposed on the mold. Finally, the biopolymer-printed elements are removed using a heat source. This technique of holding together two devices – the printed mold and a sand pressure-relief system – has proved its effectiveness in obtaining complex shapes, but there are currently no large-scale demonstrators in the construction sector.

Interestingly, other materials can also be used as temporary supports, such as folded paper (Lloret-Fritschi et al. 2022) or printed raw earth (Bruce et al. 2022).

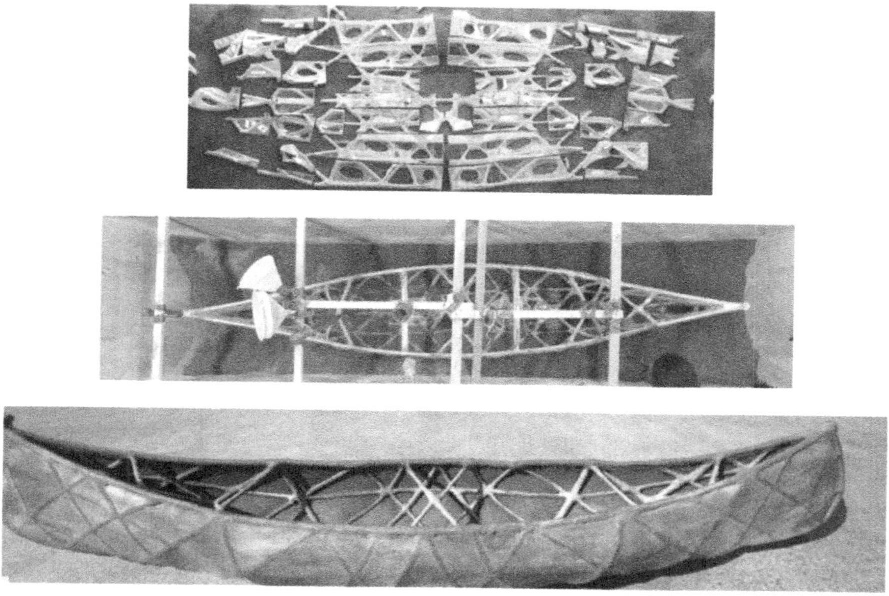

Figure 4.7. *Manufacturing a canoe by injecting cementitious material into printed biopolymer modules. Photos from Jipa and Dillenburger (2022) and Lloret-Fritschi et al. (2017) and the work of Jipa et al. (2019). Left: from top to bottom: printed parts before assembly; layout on sand; example of detail obtained. Right: printed canoe*

4.2.3.3. *Binder-jet printed supports*

The binder jetting technique can be used to print on different types of materials (geopolymer, cement, etc.). This technique consists of producing a structure by jetting a liquid binder onto a powder matrix, layer after layer. Moulds are printed and then used as formwork to place concrete, sometimes reinforced with fibers. On a large scale, we can cite several works (Ranaudo et al. 2010; Jipa et al. 2016; Aghaei Meibodi et al. 2018; Anton et al. 2020), including the beam created within the DFab and the Swiss pavilion temporarily installed in Venice in 2016 according to the thesis of Popescu (2019). Both projects combine a mold printed with this technique with shotcrete (15–20 mm thick) and cast concrete (Jipa et al. 2023). Treatment of the printed form with polyester resins is sometimes necessary to obtain a smooth surface prior to the concrete shotcrete stage.

Structures printed with these techniques are generally discretized into several parts. It is therefore necessary to plan the connections between each part (see Figure 4.8, right). The elements are then assembled by post-tensioning. Four further steps are required: mold preparation, filling by casting or spraying, concrete curing and demolding. We note that the hybrid solution of spraying concrete onto the mold to prevent cracking, then pouring the concrete to encapsulate the reinforcement, improves manufacturing time.

This process makes it possible to obtain optimized shapes and reproduce details that are impossible to reproduce using conventional manufacturing techniques and extrusion printing. The addition of shotcrete to the mold also makes it possible to produce finer, functional and aesthetic structures that would be impossible to achieve by pouring concrete. The main advantage and paradigm of 3D printing can be fulfilled by reducing the quantity of materials used. For example, the "smart slab" beam has made it possible to reduce the weight of the concrete used by 40% compared with a prestressed lightweight concrete beam (Jipa et al. 2023). However, this process is still time consuming, particularly during the design/discretization of the elements and the manufacturing/assembly stage. As an example, the beam shown in Figure 4.8 required 181 printed parts, 114 polymer printed connectors and 2,140 plywood panels. It is also necessary to master several complementary techniques, digital or otherwise, for assembly (3D plastic printing, laser cutting, milling, etc.).

Figure 4.8. *Left: 3D printed mold; spraying concrete onto the mold. Right: assembling the structure; photographs taken from Jipa et al. (2023)*

4.2.3.4. *Injection into a flexible support*

In their fresh state, cementitious materials are subject to Newton's first law, and all the problems encountered during extrusion stem from the force of gravity. In order to counteract gravity and provide support for the concrete as it is deposited, it is possible to immerse the robot's needle or nozzle in a carrier fluid and place the cementitious material in it (Benamara et al. 2020). The injection technique therefore involves extruding a fluid cementitious material into another fluid material with specific rheological properties. Archimedes' principle is not sufficient to hold the cementitious material. The carrier fluid acts as a temporary support until the element solidifies if it has yield stress fluid behavior. As a reminder, this type of fluid behaves like a solid below a certain shear stress and like a fluid above it. Once the cementitious material has been placed, the carrier fluid must therefore have a high yield stress to counteract the drag force of the deposited material. It must also be thixotropic so that it returns to its initial state as quickly as possible after passing through the nozzle.

More widely used in the medical field for the production of cells (Murphy and Atala 2014; Eshkalak et al. 2020), this strategy guarantees precision and efficiency with regard to the risk of the structure collapsing during printing. Injection and then the stability of the printed form also depend on rheological and physicochemical compatibility. It is necessary to avoid the formation of co-products through any interaction between the injected material and the carrier liquid. The recent study by Lowke et al. (2021) provides an overview of some of the physics underlying these issues. Figure 4.9 shows some examples of parts produced using this technique. The transition to higher printable volumes still depends on the use of larger containers.

Figure 4.9. *Illustrations of the 3D injection printing process. From left to right: injection process in mineral fluid; shape printed in gel; final shape after extraction (photographs taken from Hack et al. 2020, licensed under CC BY 4.0)*

4.3. Particle bed methods

The use of selective binding/particle activation technology, which is mainly used in other industries, is more dedicated to niche applications (Lowke et al. 2020;

Buswell et al. 2022). The principle behind the selective binding/activation technique is simple: a fluid material is deposited by the nozzle of a 3D printer, which then penetrates a bed of solid particles to bind them together (Lowke et al. 2022). These steps are then repeated layer by layer until a three-dimensional structure is obtained (Lowke et al. 2018, 2020). The selective bonding technique uses a mineral suspension, while the activation technique involves injecting or depositing an activating solution onto a reactive mineral powder mixed with aggregates.

For each technique, the chronology is similar: powder application, powder smoothing or compaction, fluid deposition and spreading/penetration, removal of non-bonded particles and curing of the printed part.

The success of a 3D printing process (illustrated in Figure 4.10) using selective bonding or activation depends on the control of several parameters: the characteristics of the powder (shape, porosity, roughness, etc.) and those of the binder (liquid volume, rheological properties, surface tension, etc.). Certain disadvantages arise with selective bonding techniques, such as increased porosity of the printed element and lower mechanical performance. In addition, liquid diffusion within the particle bed and post-production processing remains a challenge (Mai et al. 2022).

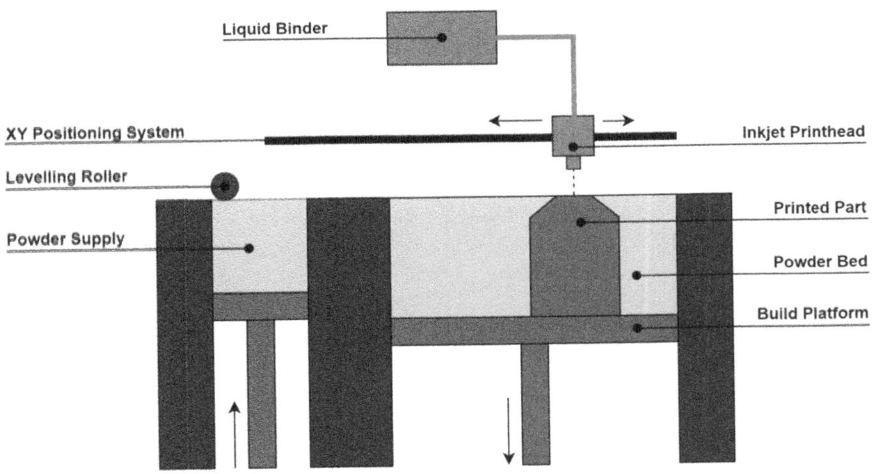

Figure 4.10. *Illustration of powder bed 3D printing process (© FM1418, w.wiki/BUQm, licensed under CC BY-SA 4.0)*

The particle sizes used with the selective powder activation technique are generally smaller than a millimeter. With the selective aggregate bonding technique, the size range is slightly larger, up to a centimeter. A suspension of cementitious materials, with complex fluid rheological behavior, is then injected. When larger aggregates are used, the suspension is less fluid (higher yield stress) and the deposited layer is then lightly compacted to ensure the bond between each layer of particles.

For both techniques, the objective is to see the porous medium completely filled by the injected or deposited liquid. In the case of cement activation, this is crucial to enable the dissolution reactions and subsequent hydration of the cement grain particles. For the selective bonding technique, if the porosity is too low or the liquid not fluid enough to flow well into the interstices, then the mechanical properties and bond life of the element will be reduced. However, if the suspension is too liquid, then the precision of the final geometry cannot be achieved, as particle bonding will not be mastered. Similarly, if the liquid diffuses throughout the porosity rather than locally, cement activation will be affected (partial hydration phenomenon). The physicochemical phenomena associated with these two techniques are beginning to be documented in the literature, for both the activation and selective bonding techniques (Pierre et al. 2018, 2020; Lowke et al. 2020, 2022; Weger et al. 2021; Zuo et al. 2022). We note that the selective bonding technique has, at present, made it possible to integrate a structural reinforcement technique using 3D metal fusion printing during the process (Straßer et al. 2024). Finally, the previous edition recalls the main phenomena brought into play during these two processes (Perrot 2019).

4.4. Support-free methods

Shotcrete is a technique that involves pressurizing concrete and then placing it spatially. This technique was invented at the beginning of the 20th century and has been used since the 1930s, notably for tunnels (Bedarf et al. 2021). The adaptation of shotcrete to a permeable fiberglass form, following a robotic trajectory, was then developed in 2019 (Taha et al. 2019).

At the end of the robot, material ejection is accelerated by an air compressor, and setting accelerators are also added to the compressed air stream. The advantage is to increase the contact surface between the air and the concrete to improve the effectiveness of the additives. Three main parameters need to be controlled: the volume of material output, the volume of compressed air introduced and the dosage of additives (setting accelerators). The 3D-printed shotcrete technique offers a wide range of ways of depositing the material. The robotic parameters are freer than with conventional extrusion: speed of travel, distance from the support, nozzle diameter, angle used for projection, etc. Researchers have linked these parameters to the properties of the final geometry by establishing empirical models (Lachmayer et al. 2023).

The advantage of this technique is to increase reinforcement possibilities compared to the traditional extrusion technique (Lindemann et al. 2019). Recent work by a research team based in Germany (Kloft et al. 2020) has shown that the kinetic energy of concrete spraying increases the density of the elements. Compared with a conventional extrusion process, mechanical properties in compression and flexion are de facto improved. Finally, the interface between each layer of the material seems to be better than with extrusion. Using X-ray tomography tests, they observed a lower quantity of trapped air on spray-printed specimens than on extrusion-printed specimens. However, precise control of the deposited geometry still needs to be improved, due to the significant kinetic effects during deposition. Figure 4.11 shows examples of 3D spray-printed concrete.

Figure 4.11. *Examples of 3D-printed projects using shotcrete. Left: spraying on textile fibers; right: structure of various cross-sections (photographs from Lowke et al. 2024, licensed under CC BY 4.0)*

4.5. Choice of process

The techniques presented in this chapter offer variable specificities in terms of resolution and printable element size. In Figure 4.12, the techniques are compared according to the essential properties that condition their choice for producing the element. Figures 4.12 (a) and (b) show, respectively, two classifications of techniques representing freedom of form as a function of substrate and resolution as a function of printed volume. The data are taken from the literature presented in this chapter.

As each technique has its own particularities, we have also grouped the advantages and limitations of each type of formwork used to produce an element in Table 4.1.

Figure 4.12. *Choice of process: (a) freedom of form according to substrate; (b) resolution according to volume*

Formwork	Alternative technique	Advantages	Limitations
Non	Shotcrete	Fast Interlayer interface Mechanical strength Reinforcement	Resolution Adhesion Spatial control of final geometry Freedom of form
Granular and Temporary	Selective bonding/particle bed technology Binder jetting	Resolution Accuracy Shape complexity Spatial control of final geometry Anti-gravity process No internal stresses Porous/lightweight structures	Heterogeneous penetration and hydration Mechanical resistance Anisotropy Reinforcement Time
Permeable, lost and permanent	Mesh Mold, Knitcrete, aerocrete, eggshell	Final geometry Spatial control of final geometry Fibers	Robot synchronization (reinforcement and concrete) Rheology and mesh size Gravity
In motion and punctual	Smart dynamic casting	Structure sizing Spatial control and surface appearance Fibers Metal reinforcement Form enhancement	Stress or deformation imposed on the material Friction Suitability for hydration Weight of formwork on articulated arm
Fluid and temporary support	Injection	Lightweight structures Shape complexity Anti-gravity process	Suitability of injected fluid and carrier fluid Inter-layer (presence of carrier fluid) Physicochemical compatibility of the two fluids Post-treatment

Table 4.1. *Advantages and limitations of alternative techniques, depending on the formwork concept*

4.6. Prospects and opportunities

4.6.1. *Dissolvable temporary supports*

The advantage of solid support is that it can be re-used, but this is in contradiction with the advantage of 3D printing's freedom of form. We may well question the relevance of creating modules that are repeated. Among these solid supports, ideas favoring printed formwork that fades with time or external stress are interesting (Jipa and Dillenburger 2022). These include dissolvable clay or polymer formwork. Figure 4.13 illustrates projects based on this technical solution.

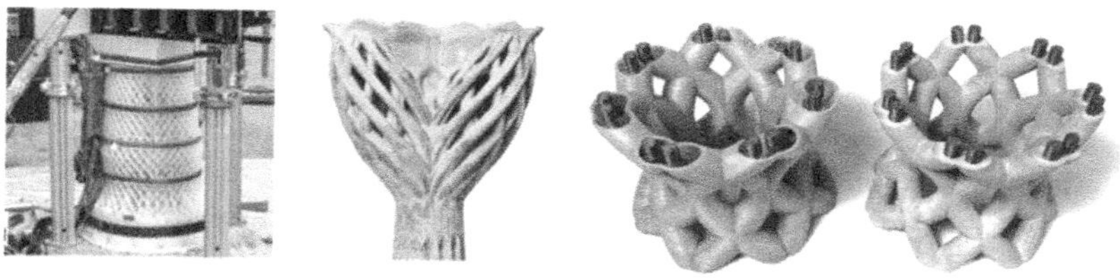

a) b)

Figure 4.13. *(a) Dissolvable formwork and final column developed at ETH Zurich and (b) example using technology developed at Iowa State University. Figures from Jipa and Dillenburger (2022). Photo credit (a) DBT, ETH Zürich; photo credit (b) Shelby Elizabeth Doyle/Iowa State University*

4.6.2. *Selective binding of aggregates with foams*

The construction industry's ecological objectives call for the use of less cement, and in general, less mineral binder in material formulations. An interesting challenge is to advance our knowledge of the behavior of mineral foams. Indeed, work has shown that it is possible to envisage inter-particle bonding using mineral foams with a very low amount of binder (Pitois et al. 2023).

4.6.3. *Structural reinforcements*

The reinforcement of printed concrete structures can be achieved using alternative printing techniques. As a reminder, the tensile strength of concrete is generally negligible, and the strength of the structure is supported by a reinforcement made of another material (steel, carbon, polymer, etc.). To date, no additive manufacturing technology has been able to compete with traditional construction techniques in building effective, low-cost reinforced elements.

Several strategies have yet to be developed alongside the use and development of alternative 3D printing techniques (Lloret-Fritschi et al. 2020).

The challenge for mass integration of 3D printing, apart from the financial cost, is the appropriate reinforcement method for any 3D printing technique. The post-tensioning system remains an effective method of joining elements without compromising all the advantages of printed concrete. For these alternative techniques, three strategies are possible: placing the metal reinforcement in either an "active" or "passive" role before the concrete is poured, or during production (either simultaneously or with fibrous reinforcements included in the formulation), or after the process (post-tensioning). In the case of passive reinforcement, the metal bars must be sufficiently bent to match the final shape and be embedded in the cementitious material.

Alternative techniques are suitable for creating honeycomb shapes, but also with reservations (voids in the structure), where ducts can pass through. In particular, these ducts can be used for post-tensioning reinforcement. These assemblies or post-tensioning reinforcements are fairly simple and suitable for fairly large scales (at least 1 m). The challenge for the future is to integrate these post-printing reinforcements with a high level of detail and a simpler steel anchoring system than for structures on a scale of several meters. Finally, when it comes to reinforcing the material, the use of fibers is an ideal solution. Although materials incorporating structural fibers lack the ductility and tensile strength of conventional metal reinforcements, they do not constrain the geometry of the part. On the other hand, they are currently scarcely used in particle-bonding techniques, but represent a real opportunity for this technique. Recent work shows that simultaneous reinforcement by melting and then welding of steel can be used to form reinforcements of variable diameter and curved structures (Mechtcherine et al. 2018; Weger et al. 2020; Straßer et al. 2022). Current work suggests that, depending on the use of the element in the building or structure, reinforcement may ultimately be a combination of different strategies (fibers + post-tensioning in particular).

4.6.4. *Functionally graded materials*

Some of the techniques presented offer shaping possibilities where one or more material properties can vary spatially. The performance of a material can indeed be driven by a gradient of properties which can be mechanical, thermal or hydric (Oxman et al. 2011; Keating and Oxman 2013; El-Galy et al. 2019). Recent work is paving the way for the development of improved thermomechanical properties of lightweight structures in these materials (Craveiro et al. 2020). An example of this is the production of functionally graded concrete using cork particles arranged gradually within the material. In particular, this work shows that a gradation of cork with a decreasing quantity from the outside to the inside of the structure requires less energy to maintain the comfort temperature. An alternative technique for achieving functional gradation within the material appears to be already in use with foams. Indeed, graduated porosities are included in concretes by incorporating different quantities of foaming agents (Oxman et al. 2011). These studies pave the way for exploiting alternative techniques for greater functionality in buildings.

4.7. Conclusion

The new techniques described in this chapter offer innovative approaches to design, production and assembly. Most of them, however, appear to be better suited to off-site construction. In view of recent developments in site logistics optimization, it would be interesting to focus future studies on these alternative techniques. Indeed, the alternative techniques that have emerged in recent years are well suited to prefabrication. They are currently complementary to traditional construction techniques, but could play an increasingly important role in the years to come, particularly by integrating circularity and the reuse of end-of-life components. Increasing the use of these techniques will be possible provided that scientific barriers are removed and their advantages reinforced. It is also worth pointing out that the use of these techniques is only possible with qualified workers and the emergence of 3D printing in education.

4.8. References

Aghaei Meibodi, M., Jipa, A., Giesecke, R., Shammas, D., Bernhard, M., Leschok, M., Graser, K., Dillenburger, B. (2018). Smart slab: Computational design and digital fabrication of a lightweight concrete slab. In *Acadia 2018 Recalibration: On Imprecision and Infidelity: Proceedings of the 38th Annual Conference of the Association for Computer Aided Design in Architecture*, Anzalone, P., del Signore, M., Wit, A.J. (eds). Association for Computer Aided Design in Architecture (ACADIA), Mexico City.

Anton, A., Jipa, A., Reiter, L., Dillenburger, B. (2020). Fast complexity: Additive manufacturing for prefabricated concrete slabs. In *Second RILEM International Conference on Concrete and Digital Fabrication: Digital Concrete 2020*. Springer, Heidelberg.

Bedarf, P., Dutto, A., Zanini, M., Dillenburger, B. (2021). Foam 3D printing for construction: A review of applications, materials, and processes. *Automation in Construction*, 130, 103861.

Bedarf, P., Calvo-Barentin, C., Schulte, D.M., Şenol, A., Jeoffroy, E., Dillenburger, B. (2023). Mineral composites: Stay-in-place formwork for concrete using foam 3D printing. *Archit. Struct. Constr.*, 3, 251–262. doi: 10.1007/s44150-023-00084-x.

Benamara, A., Pierre, A., Kaci, A., Melinge, Y. (2020). 3D printing of a cement-based mortar in a complex fluid suspension: Analytical modeling and experimental tests. In *Second RILEM International Conference on Concrete and Digital Fabrication: Digital Concrete 2020*. Springer, Heidelberg.

Boyd IV, R.P., Weller, C., DiSanto, A., Rees, M., Hilbert, B. (2020). Cellular fabrication and apparatus for additive manufacturing. Google Patents, US 2017/0217088 A1.

Bruce, M., Clune, G., Xie, R., Mozaffari, S., Adel, A. (2022). Cocoon: 3D printed clay formwork for concrete casting. In *Annual Conference of the Association of Computer Aided Design in Architecture, CUMINCAD*, Cairo, 400–409. doi: 10.7302/7046.

Buswell, R.A., Bos, F.P., da Silva, W.R.L. Hack, N., Kloft, H., Lowke, D., Freund, N., Fromm, A., Dini, E., Wangler, T. (2022). Digital fabrication with cement-based materials: Process classification and case studies. In *Digital Fabrication with Cement-Based Materials: State-of-the-Art Report of the RILEM TC 276-DFC*, Roussel, N. and Lowke, D. (eds). Springer, Cham, 11–48.

Craveiro, F., Nazarian, S., Bartolo, H., Bartolo, P.J., Duarte, J.P. (2020). An automated system for 3D printing functionally graded concrete-based materials. *Additive Manufacturing*, 33, 101146.

El-Galy, I.M., Saleh, B.I., Ahmed, M.H. (2019). Functionally graded materials classifications and development trends from industrial point of view. *SN Applied Sciences*, 1(11), 1378.

Eshkalak, S.K., Ghomi, E.R., Dai, Y., Choudhury, D., Ramakrishna, S. (2020). The role of three-dimensional printing in healthcare and medicine. *Materials & Design*, 194, 108940.

Goyon, J., Bertrand, F., Pitois, O., Ovarlez, G. (2010). Shear induced drainage in foamy yield-stress fluids. *Physical Review Letters*, 104(12), 128301.

Hack, N. and Lauer, W.V. (2014). Mesh-mould: Robotically fabricated spatial meshes as reinforced concrete formwork. *Architectural Design*, 84(3), 44–53.

Hack, N., Dressler, I., Brohmann, L., Gantner, S., Lowke, D., Kloft, H. (2020). Injection 3D Concrete Printing (I3DCP): Basic principles and case studies. *Materials*, 13(5). doi: 10.3390/ma13051093.

Jipa, A. and Dillenburger, B. (2022). 3D printed formwork for concrete: State-of-the-art, opportunities, challenges, and applications. *3D Printing and Additive Manufacturing*, 9(2), 84–107.

Jipa, A., Bernhard, M., Meibodi, M., Dillenburger, B. (2016). 3D-printed stay-in-place formwork for topologically optimized concrete slabs. In *Proceedings of the 2016 TxA Emerging Design+ Technology Conference*. Texas Society of Architects, San Antonio, Texas. doi: 10.3929/ethz-b000237082.

Jipa, A., Bernhard, M., Ruffray, N., Wangler, T., Flatt, R.J., Dillenburger, B. (2019). Formwork fabrication freedom for a concrete canoe. *Gestão & Tecnologia de Projetos*, 14(1), 25–44.

Jipa, A., Bürgin, T., Dillenburger, B. (2023). 3D-printed formworks for glass fibre reinforced shotcrete. In *GRCA 2023*.

Keating, S. and Oxman, N. (2013). Compound fabrication: A multi-functional robotic platform for digital design and fabrication. *Robotics and Computer-Integrated Manufacturing*, 29(6), 439–448.

Kloft, H., Krauss, H.-W., Hack, N., Herrmann, E., Neudecker, S., Varady, P.A., Lowke, D. (2020). Influence of process parameters on the interlayer bond strength of concrete elements additive manufactured by Shotcrete 3D Printing (SC3DP). *Cement and Concrete Research*, 134, 106078.

Kuzmenko, K., Ducoulombier, N., Feraille, A., Roussel, N. (2022). Environmental impact of extrusion-based additive manufacturing: Generic model, power measurements and influence of printing resolution. *Cement and Concrete Research*, 157, 106807.

Lachmayer, L., Böhler, D., Freund, N., Mai, I., Lowke, D., Raatz, A. (2023). Modelling the influence of material and process parameters on Shotcrete 3D Printed strands-cross-section adjustment for automatic robotic manufacturing. *Automation in Construction*, 145, 104626.

Lindemann, H., Gerbers, R., Ibrahim, S., Dietrich, F., Herrmann, E., Dröder, K., Raatz, A., Kloft, H. (2019). Development of a shotcrete 3D-printing (SC3DP) technology for additive manufacturing of reinforced freeform concrete structures. In *First RILEM International Conference on Concrete and Digital Fabrication–Digital Concrete 2018*. Springer, Heidelberg.

Lloret-Fritschi, E., Reiter, L., Wangler, T., Gramazio, F., Kohler, M., Flatt, R.J. (2017). Smart dynamic casting: Slipforming with flexible formwork-inline measurement and control. In *HPC/CIC Tromsø 2017*, 27.

Lloret-Fritschi, E., Wangler, T., Gebhard, L., Mata-Falcón, J., Mantellato, S., Scotto, F., Burger, J., Szabo, A., Ruffray, N., Reiter, L. (2020). From smart dynamic casting to a growing family of digital casting systems. *Cement and Concrete Research*, 134, 106071.

Lloret-Fritschi, E., Choma, J., Scotto, F., Szabo, A., Gramazio, F., Kohler, M., Flatt, R.J. (2022). In-crease: Less concrete more paper. *RILEM Technical Letters*, 7, 199–208.

Lowke, D., Dini, E., Perrot, A., Weger, D., Gehlen, C., Dillenburger, B. (2018). Particle-bed 3D printing in concrete construction – Possibilities and challenges. *Cement and Concrete Research*, 112, 50–65.

Lowke, D., Talke, D., Dressler, I., Weger, D., Gehlen, C., Ostertag, C., Rael, R. (2020). Particle bed 3D printing by selective cement activation – Applications, material and process technology. *Cement and Concrete Research*, 134, 106077.

Lowke, D., Vandenberg, A., Pierre, A., Thomas, A., Kloft, H., Hack, N. (2021). Injection 3D concrete printing in a carrier liquid – Underlying physics and applications to lightweight space frame structures. *Cement and Concrete Composites*, 124, 104169.

Lowke, D., Mai, I., Keita, E., Perrot, A., Weger, D., Gehlen, C., Herding, F., Zuo, W., Roussel, N. (2022). Material-process interactions in particle bed 3D printing and the underlying physics. *Cement and Concrete Research*, 156, 106748.

Lowke, D., Anton, A., Buswell, R., Jenny, S.E., Flatt, R.J., Fritschi, E.L., Hack, N., Mai, I., Popescu, M., Kloft, H. (2024). Digital fabrication with concrete beyond horizontal planar layers. *Cement and Concrete Research*, 186, 107663. doi: 10.1016/j.cemconres.2024.107663.

Mai, I., Lowke, D., Perrot, A. (2022). Fluid intrusion in powder beds for selective cement activation – An experimental and analytical study. *Cement and Concrete Research*, 156, 106771.

Mechtcherine, V., Grafe, J., Nerella, V.N., Spaniol, E., Hertel, M., Füssel, U. (2018). 3D-printed steel reinforcement for digital concrete construction – Manufacture, mechanical properties and bond behaviour. *Construction and Building Materials*, 179, 125–137.

Murphy, S.V. and Atala, A. (2014). 3D bioprinting of tissues and organs. *Nature Biotechnology*, 32(8), 773–785.

Oxman, N., Keating, S., Tsai, E. (2011). Functionally graded rapid prototyping. *Innovative Developments in Virtual and Physical Prototyping*, 483–489.

Perrot, A. (2019). *3D Printing of Concrete: State of the Art and Challenges of the Digital Construction Revolution*. ISTE Ltd, London, and John Wiley & Sons, New York.

Perrot, A., Pierre, A., Nerella, V.N., Wolfs, R.J.M., Keita, E., Nair, S.A.O., Neithalath, N., Roussel, N., Mechtcherine, V. (2021). From analytical methods to numerical simulations: A process engineering toolbox for 3D concrete printing. *Cement and Concrete Composites*, 122, 104164.

Pierre, A., Weger, D., Perrot, A., Lowke, D. (2018). Penetration of cement pastes into sand packings during 3D printing: Analytical and experimental study. *Materials and Structures*, 51, 1–12.

Pierre, A., Weger, D., Perrot, A., Lowke, D. (2020). 2D numerical modelling of particle-bed 3D printing by selective paste intrusion. In *Second RILEM International Conference on Concrete and Digital Fabrication: Digital Concrete 2020*. Springer, Heidelberg.

Pitois, O., Salamé, A., Khidas, Y., Ceccaldi, M., Langlois, V., Vincent-Bonnieu, S. (2023). Daisy-shaped liquid bridges in foam-filled granular packings. *Journal of Colloid and Interface Science*, 638, 552–560.

Popescu, M.A. (2019). KnitCrete: Stay-in-place knitted formworks for complex concrete structures. PhD Thesis, ETH Zurich.

Popescu, M.A., Reiter, L., Liew, A., Van Mele, T., Flatt, R.J., Block, P. (2018). Building in concrete with an ultra-lightweight knitted stay-in-place formwork: Prototype of a concrete shell bridge. *Structures*, 14, 322–332. doi: 10.1016/j.istruc.2018.03.001.

Ranaudo, F., Van Mele, T., Block, P. (2010). A low-carbon, funicular concrete floor system: Design and engineering of the HiLo floors. *System*, 2012(2016), 2016–2018.

Rennen, P., Gantner, S., Dielemans, G., Bleker, L., Christidi, N., Dörrie, R., Hojjat, M., Mai, I., Mawas, K., Lowke, D. et al. (2023). Robotic knitcrete: Computational design and fabrication of a pedestrian bridge using robotic shotcrete on a 3DKnitted formwork. *Front. Built Environ.*, 9, 1269000. doi: 10.3389/fbuil.2023.1269000.

Straßer, A., Weger, D., Matthäus, C., Kränkel, T., Gehlen, C. (2022). Combining wire and arc additive manufacturing and selective paste intrusion for additively manufactured structural concrete: Fundamental investigations on the effect of heat exposure by WAAM on the rheological and intrusion behavior of cement paste in the particle bed during selective paste intrusion. In *Open Conference Proceedings*, 1, 61–72. doi: 10.52825/ocp.v1i.75.

Straßer, A., Haynack, A., Kränkel, T., Gehlen, C. (2024). Additive manufacturing by the selective paste intrusion: Effect of the distance of the print nozzle to the particle bed on the print quality. *Construction and Building Materials*, 449, 138274. doi: 10.1016/j.conbuildmat.2024.138274.

Szabo, A. (2020). Design and fabrication of thin folded members with digital concrete processes. PhD Thesis, ETH Zurich.

Szabo, A., Lloret-Fritschi, E., Reiter, L., Gramazio, F., Kohler, M., Flatt, R.J. (2019a). Revisiting folded forms with digital fabrication. In *Proceedings of the eCAADe SIGraDi 2019, Architecture in the Age of the 4th Industrial Revolution*, Sousa, J.P., Castro Henriques, G., Xavier, J.P. (eds). FAUP, Porto.

Szabo, A., Reiter, L., Lloret-Fritschi, E., Gramazio, F., Kohler, M., Flatt, R.J. (2019b). Adapting smart dynamic casting to thin folded geometries. In *First RILEM International Conference on Concrete and Digital Fabrication–Digital Concrete 2018*. Springer, Heidelberg.

Szabo, A., Reiter, L., Lloret-Fritschi, E., Wangler, T., Gramazio, F., Kohler, M., Flatt, R.J. (2020a). ACDC: The admixture controlled digital casting and its application to thin folded concrete structures. In *Second RILEM International Conference on Concrete and Digital Fabrication: Digital Concrete 2020*. Springer, Cham. doi: 10.1007/978-3-030-49916-7_93.

Szabo, A., Reiter, L., Lloret-Fritschi, E., Gramazio, F., Kohler, M., Flatt, R.J. (2020b). Mastering yield stress evolution and formwork friction for smart dynamic casting. *Materials*, 13(9), 2084. doi: 10.3390/ma13092084.

Taha, N., Walzer, A.N., Ruangjun, J., Bürgin, T., Dörfler, K., Lloret-Fritschi, E., Gramazio, F., Kohler, M. (2019). Robotic AeroCrete – A novel robotic spraying and surface treatment technology for the production of slender reinforced concrete elements. In *Architecture in the Age of the 4th Industrial Revolution–Proceedings of the 37th eCAADe and 23rd SIGraDi Conference*. CumInCAD, Porto.

Vasilic, K., Meng, B., Kühne, H.-C., Roussel, N. (2011). Flow of fresh concrete through steel bars: A porous medium analogy. *Cement and Concrete Research*, 41(5), 496–503.

Vasilic, K., Schmidt, W., Kühne, H.-C., Haamkens, F., Mechtcherine, V., Roussel, N. (2016). Flow of fresh concrete through reinforced elements: Experimental validation of the porous analogy numerical method. *Cement and Concrete Research*, 88, 1–6.

Weger, D., Baier, D., Straßer, A., Prottung, S., Kränkel, T., Bachmann, A., Gehlen, C., Zäh, M. (2020). Reinforced particle-bed printing by combination of the selective paste intrusion method with wire and arc additive manufacturing – A first feasibility study. In *Second RILEM International Conference on Concrete and Digital Fabrication: Digital Concrete 2020*. Springer, Cham.

Weger, D., Pierre, A., Perrot, A., Kränkel, T., Lowke, D., Gehlen, C. (2021). Penetration of cement pastes into particle-beds: A comparison of penetration models. *Materials*, 14(2), 389.

Wu, P., Wang, J., Wang, X. (2016). A critical review of the use of 3-D printing in the construction industry. *Automation in Construction*, 68, 21–31.

Zuo, W., Dong, C., Belin, P., Roussel, N., Keita, E. (2022). Capillary imbibition depth in particle-bed 3D printing – Physical frame and one-dimensional experiments. *Cement and Concrete Research*, 156, 106740.

5

Structural Applications of 3D Printing

Romain MESNIL[1], Romain DUBALLET[2] and Olivier BAVEREL[1]

[1]*Laboratoire Navier, École des Ponts, Marne-la-Vallée, France*

[2]*XTreeE, Rungis, France*

5.1. Introduction

5.1.1. *Structure design*

Construction materials are responsible for over 7% of man-made greenhouse gas emissions and account for almost half of humanity's use of raw materials. Structural designers are therefore obliged to couple material savings with the safety of users of future infrastructures (Baverel et al. 2022). René Motro has identified five structural design levers available to engineers and architects in a design approach known as *structural morphology*:

– The *material*.

– The *form*, which influences the distribution of forces in the structure.

– The *structure*, which is the way elements are arranged together.

– The *forces*, which are the external or internal forces applied to the structure.

– The *technology*, which encompasses all technical solutions for the manufacture and assembly of structural elements.

For a color version of all figures in this chapter, see www.iste.co.uk/perrot/3dconcrete.zip.

3D Concrete Printing,

In the case of 3D concrete printing, concrete remains a material that performs well in compression and poorly in tension, whatever the chosen material mix-design. This leaves engineers with the forces, the technology, the structure and the shape. The latter is an obvious lever for structural optimization, since it is well known that an arch uses the material much more efficiently than a beam in bending. 3D printing offers greater formal freedom than traditional construction processes, and therefore seems to lead to more economical designs in terms of building materials. However, this intuition is largely just a promise yet to be fulfilled.

The notion of arrangement between components with hierarchical relationships remains essential to design structures and makes them readable. This compositional work remains relatively in the background of current 3D printing research. However, it is one of the key issues in the effective deployment of this technology and has historically been a difficult barrier to overcome. It is interesting to note that modern building materials have led to structural systems that benefit the most from their mechanical properties. Early metal structures, such as the Iron Bridge designed by Thomas Farnolls Pritchard in 1779 are based on the vocabulary of masonry structures, while the truss bridges built by Gustave Eiffel take advantage of metal's excellent tensile strength. Similarly, the use of prestressing enabled Eugène Freyssinet to take advantage of concrete's high compressive strength more than a century after Vicat created modern concretes in 1817. More recently, composite materials, which are characterized by high deformation at failure, have led to explorations of elastic gridshells, several decades after their creation (Caron et al. 2018).

(a) Iron Bridge (1779): a structure inspired by stone voussoirs. Image credit: Nilfanion, distributed under CC BY-SA 4.0 license.

(b) Viaduc de Garabit (1884): the triumph of space trusses. Image credit: Patrick Giraud, distributed under CC BY-SA 3.0 license.

Figure 5.1. *Evolution of the structural morphology of steel bridges*

These few examples illustrate the need to consider the construction system in conjunction with the technology used to shape the material. It is thus likely that truly relevant printed structures will not be simple adaptations of traditional reinforced concrete structures. The difficulty of this approach becomes clear when we consider the time it takes from the creation of a material (or process) to the construction of structural masterpieces that synthesize all the dimensions of structural morphology.

5.1.2. *Principles of structural design*

Engineering is the art of shaping materials we do not fully understand, into forms we cannot accurately analyze in order to withstand forces we cannot properly assess, in such a way that the public has no reason to suspect the extent of our ignorance.

This definition by Archie R. Dykes, in a speech to the British Institution of Structural Engineers in 1976, shows that structural engineers have to design under uncertainty. A probability-based approach to design was developed in the second half of the 20th century, then formalized, notably with the Eurocodes. The Eurocodes dimensioning approach is based on the principle of *limit states* and distinguishes between two categories:

– *The ultimate limit state* (ULS) verifies that the applied loads do not lead to structural failure.

– *The serviceability limit state* (SLS) verifies that the most common loads do not lead to a loss of the element's functionality.

The loads applied to the structure (self-weight G, wind W, snow S, temperature T and operating loads Q) are calculated on the basis of an estimated building life (typically 50 years). These loads are then combined, bearing in mind that the combination of several simultaneous loads is rarer than the occurrence of a single one. Of course, self-weight is always taken into account in load combinations. The loads q_{ELU} applied at the ULS are greater than the loads applied at the SLS q_{ELS}: we accept a loss of building functionality for these loads, which have a very rare probability of occurrence in the liftetime of a structure.

The design strength of the material is a function of the characteristic strength f_k, which is the strength at the $5^t h$ percentile, and the material's partial safety coefficient γ_m. In the case of concrete, $\gamma_m = 1.5$.

$$f_d = \frac{f_k}{\gamma_m} \qquad [5.1]$$

Note that two materials with the same mean strength value do not necessarily have the same characteristic value and therefore the same design strength. This is illustrated in Figure 5.2, which shows the probability densities for the failure stress of two materials. Although the mean strength is the same, material 2 (shown in ochre) has a higher characteristic strength thanks to less scatter in the results.

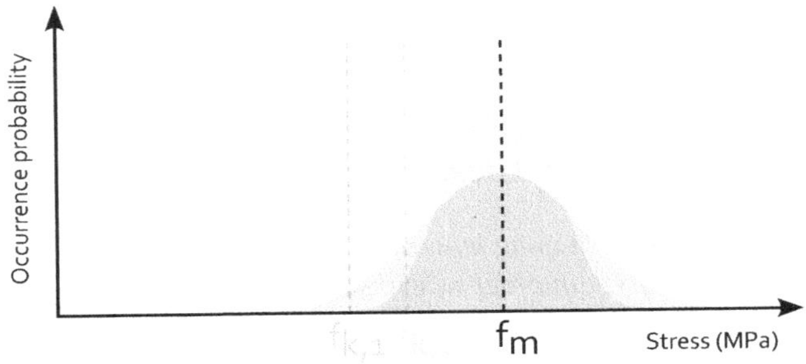

Figure 5.2. *Two materials with the same average strength, but different design strengths*

The principle of ULS design is illustrated in Figure 5.3. It shows two probability densities of stress level and material strength, weighted by safety coefficients.

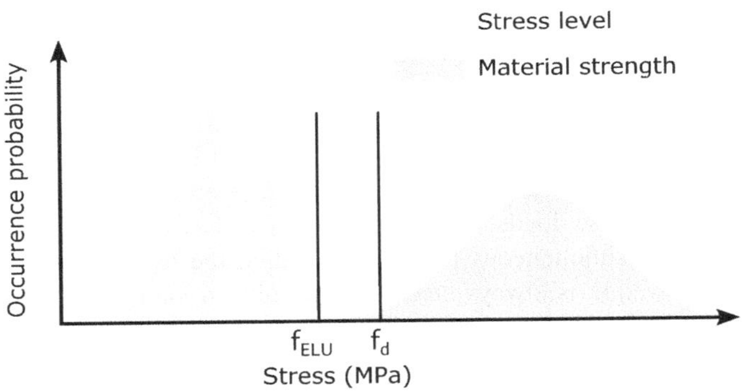

Figure 5.3. *Principle of Eurocode ultimate limit state design: a probable stress level is compared with the material's design strength*

The other main principle of the codes is ductility: we want to be able to observe disorders before a building totally (or partially) collapses. Concrete is brittle in tension, but pseudo-ductile in compression. Three main strategies have been developed to overcome this brittleness:

– the use of *non-reinforced* material in predominantly compressed structures;

– the use of concrete with reinforcement or fibers to achieve ductile behavior. Eurocode 2 requires the use of passive reinforcement in seismic zones;

– the use of (compressive) prestressing, enabling the material to be used in compression, even for bent structures.

The design of printed concrete structures will therefore have to fit into a framework of standards that is already well established. The tests will have to be easily reproducible by economic actors, and a certain ductility will have to be demonstrated.

5.1.3. *Chapter organization*

Concrete 3D printing needs to move toward structural applications to have a significant impact on the construction industry and this is the thrust of research work being carried out in several laboratories (Asprone et al. 2018; Marchment and Sanjayan 2020; Caron et al. 2021). Historical analysis shows that shapes and structures evolve with the technologies and materials available. For printed concrete structures, this evolution is not yet set in stone.

The aim of this chapter is therefore to provide an overview of the potential structural applications of printed concrete. First, it identifies mechanical models for printed structures and methods for assessing the mechanical properties of the hardened material. Second, this section presents ways of assessing the material properties required for the dimensioning process and suggests ways in which printed structures can be modeled efficiently. Once equipped with the right physical models, the engineer can explore the various dimensions of structural morphology. The third section discusses the application of printing as masonry, that is, without reinforcement, with a focus on the question of *form* and *structure*. The fourth section briefly presents applications of 3D printing in structures reinforced with reinforcement or fibers, with a focus on *technology*. The fifth section presents applications using prestressing and explores the *strength* dimension of structural morphology. The chapter concludes with a discussion of modeling and optimization tools that could be used to assist engineers in their design approach.

5.2. Assessment of hardened material properties

5.2.1. *Material or micro-structure?*

One of the first questions for an engineer calculating a structure is which model to adopt to capture the correct physical phenomena at play, and which tests to carry out to apply this model in practice. Although a volume at stake can provide accurate results, it is cumbersome and costly to implement in practice, and more difficult to interpret than a simpler model (plate or beam). Given the slenderness of the printed walls, a plate model seems more suitable, as it is compatible with structural design tools and the structural engineer's vocabulary. An immediate difficulty then arises: a 3D-printed structure has several scales to consider – that of the material, of the structure, but also that of the bead. Indeed, bead geometry can influence stress distribution in the shell. A recent study has shown that the geometry of the filaments after deposition has a significant influence on the stiffness of the structure, particularly for bending deformations (Mesnil et al. 2023). Figure 5.4 illustrates the distribution of principal stresses in a section composed of four layers subjected to uniform loading. The geometry of the filaments is inspired by the oblong model proposed by Carneau et al. (2022). We find that the major stresses have a very complex distribution, where a test on a plate of uniform thickness would have led to zero major stress and constant minor stress.

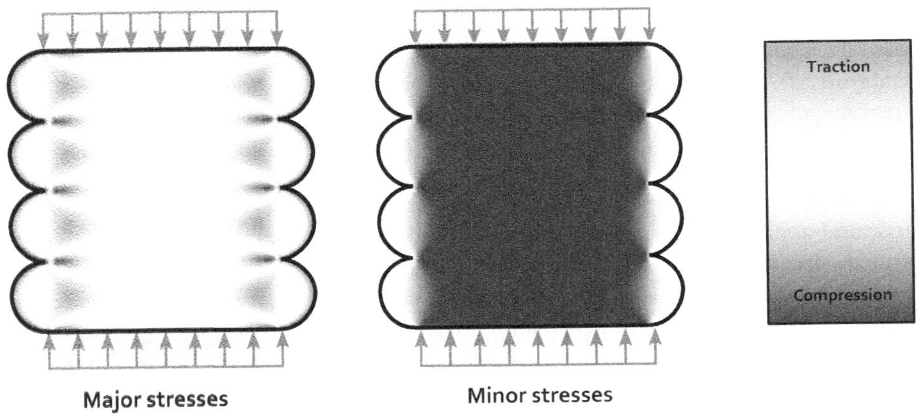

Figure 5.4. *Principal stress distributions in a printed cord under uniform load in linear elasticity*

This simple example shows that a test on a printed specimen with the protuberances resulting from printing is indeed a test of *structure* and not of *material*. This meso-structure is consubstantial with the 3D extrusion printing process. It should also be noted that, in addition to geometric heterogeneity, it is also possible to have material heterogeneity, which affects the stress distribution in a printed surface. Characterization of the printed material is therefore not necessarily the most relevant approach for obtaining input parameters for a structural calculation.

Tests on specimens printed with the real meso-structure raise new issues. For example, the width of a printed bead depends on the printing process and may differ from usual specimen widths (e.g. 40 mm for a flexural specimen for mortar).

The preparation of non-rectified specimens also raises problems for test implementation due to the brittleness of the concrete.

5.2.2. *Toward reconciling points of view*

A simpler approach than experimental characterization of the micro-structure may be to test the material properties, on the one hand, and to take into account the effect of the micro-structure by an appropriate model. Thus, it has been shown that the elastic behavior of printed structures can be described by a Kirchoff–Love plate model, with a stiffness matrix calculated by periodic homogenization (Mesnil et al. 2023). The results show that the components of this matrix are not reducible to an equivalent Young's modulus or thickness, as flexural and membrane behavior are very different. Figure 5.5 shows a comparison between a calculation with volume elements and a calculation with an equivalent plate model. The difference in deflection is of the order of 2%, with the number of vertices divided by 1,000 for the plate model.

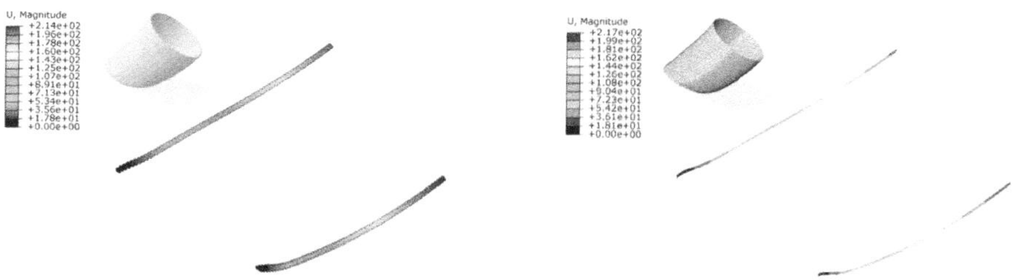

Figure 5.5. *Visualization of the deformation of an inclined cylinder with a volume model and an equivalent plate model (image: Mesnil et al. (2023))*

The advantage of this homogenization method is not only a considerable computational gain over finite volume elements, but also a better understanding of overall behavior, thanks to the use of generalized stresses and strains. In this way, the study carried out separates the influence of the material, on the one hand, and the geometry of the layers, on the other hand.

Taking account of heterogeneities in the stiffness and strength of printed structures is a recent research topic, but the characterization of printed material on geometries without asperities seems to be preferred at the moment, as it frees us from experimental complications, and enables us to transfer the know-how developed on mortar characterization. It is possible to characterize a cast material, an extruded material using laboratory equipment, or an extruded material using industrial equipment. The relative merits of the three approaches are compared in a recent publication, which serves as the basis for an international research collaboration[1] (Mechtcherine et al. 2022). While the test on cast material may yield interesting results at first glance, the influence of the process on the properties of the hardened material cannot be ignored a priori. Factors of influence include drying, mixing quality and possibly the dispersion of the setting accelerator in two-component systems.

Figure 5.6 illustrates a protocol for characterizing a printed material using this methodology:

1) print straight walls of sufficient thickness;

2) orient samples specimens in precise directions;

3) cut and surface specimens to obtain a parallelepiped geometry comparable to that used in conventional tests;

4) perform a mechanical test.

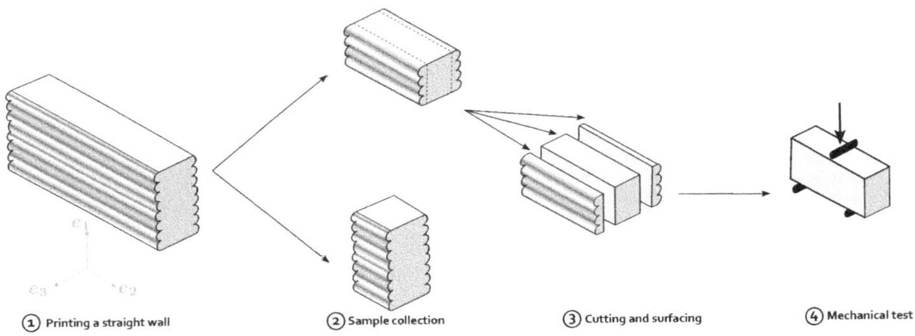

Figure 5.6. *Protocol for a printed material characterization test, the printing direction is the two-direction*

5.2.3. *Orthotropic or isotropic material?*

Structural engineers often assume that concrete is an isotropic material, with the same stiffness and strength properties regardless of the direction of loading. This assumption is not a priori directly applicable to extrusion-printed concrete, as the method of deposition can influence the material's micro-structure. Researchers have observed ellipsoidal pores elongated along the extrusion direction, as well as material orthotropy (van den Heever et al. 2022; Kruger and van der Westhuizen 2023).

1 Namely, the *Assessment of Additively Manufactured Concrete Materials and Structures* working group within RILEM.

Therefore, a priori, we must use an orthotropic material model with three directions, the extrusion direction, denoted as 2 in Figure 5.6, the transverse direction 1, perpendicular to the extrusion direction but in the mean plane of the printed wall, and the direction 3, which is perpendicular to this plane.

This potential orthotropy is the source of greater complexity in experimental characterization, which seems incompatible with practical applications. Elasticity theory tells us that an orthotropic material is characterized by nine material parameters, while an isotropic material is characterized by just two (e.g. Young's modulus and Poisson's ratio). Equation [5.2] gives the general form of the compliance matrix $\underline{\underline{S}}$, which is symmetrical: the material parameters are the moduli E_i, the shear moduli G_{ij} and the Poisson's ratio ν_{ij}. Similarly, adapting classical failure criteria such as the Drucker Prager model to the orthotropic case results in a similar increase in the number of parameters:

$$\underline{\underline{S}} = \begin{pmatrix} \frac{1}{E_1} & \frac{-\nu_{21}}{E_2} & \frac{-\nu_{31}}{E_3} & 0 & 0 & 0 \\ \frac{-\nu_{12}}{E_1} & \frac{1}{E_2} & \frac{-\nu_{32}}{E_3} & 0 & 0 & 0 \\ \frac{-\nu_{13}}{E_1} & \frac{-\nu_{23}}{E_2} & \frac{1}{E_3} & 0 & 0 & 0 \\ 0 & 0 & 0 & \frac{1}{G_{23}} & 0 & 0 \\ 0 & 0 & 0 & 0 & \frac{1}{G_{31}} & 0 \\ 0 & 0 & 0 & 0 & 0 & \frac{1}{G_{12}} \end{pmatrix} \qquad [5.2]$$

This theoretical number of parameters can be reduced in practice. Note that in many cases, 3D printing is used to produce thin-walled structures that can be modeled by plates. In such cases, the most suitable model is the plane stress model: $\sigma_{13} = \sigma_{23} = \sigma_{33} = 0$. The strength properties of the material in these load directions therefore generally do not need to be evaluated. Experiments show that Poisson's ratios do appear to vary with direction (Kruger and van der Westhuizen 2023), but this variation is small enough to be disregarded in practice, and the Poisson's ratio of the cast material can be used. Table 5.1 summarizes the properties needed to describe an orthotropic material and the relevance of testing them in practice.

Category	Property	Symbol	To be evaluated specifically
Stiffness	Young's modulus	E_i	Yes
	Shear modulus	G_{ij}	No
	Poisson's ratio	ν_{ij}	No
Robustness	Compressive strength	$f_{ck,1}$, $f_{ck,2}$	Yes
		$f_{ck,3}$	No
	Flexural strength	$f_{bk,1}$, $f_{bk,2}$	Yes
		$f_{bk,3}$	No

Table 5.1. *Summary of material properties to be evaluated*

5.2.4. *Mechanical testing*

The methodology outlined in Figure 5.6 shows that the mechanical tests carried out on printed material are identical to those performed on conventional hardened material. These include the following:

– the compression test performed on a cubic or cylindrical specimen, which is also used to assess Young's modulus;

– three-point or four-point bending test performed on a prismatic specimen;

– the uniaxial tensile test, or the splitting test (known as the "Brazilian" test).

Specimen dimensions vary according to maximum aggregate size. The dimensions specified in the standard must be respected, as concrete has a brittle fracture mode, and the measured value depends on the size of the specimen. In this respect, we refer to the work of Bažant (1999). In all cases, the experimenter must carry out sufficient tests to calculate characteristic values.

5.2.5. *Conclusion*

The 3D printing process results in a mesostructure whose mechanical properties cannot be directly assimilated to quantities commonly used by engineers (Young's modulus or Poisson's ratio). We have presented an experimental method for identifying the parameters of the hardened material by preparing sawn and rectified specimens. Using the theory of periodic homogenization, it is then possible to derive a homogeneous plate stiffness, enabling engineers to perform calculations using models they are familiar with. However, models capable of describing the ruin of a periodic structure have not been applied to the microstructure resulting from concrete printing. This means that, for the time being, we have to make do with the strengths obtained from tests on rectified specimens, until we have a better understanding of the influence of microstructure on the potential failure mechanisms of printed concrete.

Having addressed the question of the appropriate mechanical model, the remainder of this chapter will focus on the morphological strategies employed to build 3D-printed structures.

5.3. Masonry

With a view to the gradual adoption of 3D printing technologies by the industry, and although at first glance this solution is less relevant than reinforced concrete-type applications, 3D-printed structures can be designed according to Eurocode 6, that is, in masonry. There are two complementary approaches to masonry construction that differ in their strategy for placing ductile reinforcements: reinforced masonry and confined masonry.

5.3.1. *Reinforced masonry*

Reinforced masonry is a reinforcement strategy designed to place ductile elements, such as metal spacers and steel rebars, between the stacked elements of a masonry wall. It therefore implies intervention during the erection of the wall, which is quite natural for cinder block construction, but less so for 3D printing. In this case, either the process has to be interrupted to make way for human intervention, or the elements have to be placed during the printing process, either automatically or manually if the latter is slow enough to allow this.

This strategy has been used extensively for on-site 3D printing experiments, whether or not they have been the subject of normative justification. Indeed, if we take the example of on-site printing of a single house, the characteristic times between each layer deposited are sufficiently high (on the order of tens of minutes) to allow intervention on areas not stressed by the machine during printing. It should also be noted that for a variety of reasons, such as the progressive strength increase of the material, inclement weather conditions, or simply maintenance, such printing may be interrupted for periods exceeding the setting time of the concrete, thus creating vertical discontinuities.

Figure 5.7. *Metal spacers on an on-site printed wall (image: Peri SE)*

5.3.2. *Confined masonry*

Confined masonry consists of framing a wall of unreinforced material with chains, that is, longitudinal elements such as columns and beams, whose mechanical cohesion with the wall is ensured. The result is a pseudo-ductile object that can easily be used in buildings up to R+6. The advantage of this paradigm for large-scale 3D printing is that it shifts the question of reinforcement to the peripheral connections of a printed wall.

A similar principle was adopted to certify the printed walls on the south facade of the five houses in the Villiaprint project in Reims, as shown in Figure 5.8, which obtained its first Appréciation Technique d'Expérimentation (ATEx) from the CSTB in 2020 for printed walls in cementitious materials. Here, printed walls without reinforcement act as infill between reinforced concrete columns supporting a wooden roof and connected at the top by a horizontal beam. The mechanical properties required of the walls are limited to impact resistance, in-plane bracing and waterproofing.

Figure 5.8. *South facade of a house in the Villiaprint project, 2022 (XtreeE)*

From a strictly environmental and mechanical point of view, such a principle can hardly claim optimal results, given the quantities of cement-rich material involved. Rather, the interest lies in the aspect of normative experimentation, health and safety in the industrial sector (the Villiaprint project as a whole was an experiment in off-site prefabrication), as well as architectural quality (the five houses, in a social housing context, all boast unique and attractive facades, and this for virtually no extra cost, given the robotic manufacturing method). Two areas remain to be explored: the use of "low-carbon" materials with a lower impact, and the use of structure to reduce the amount of materials used.

Conventional printed walls, made of three layers of mortar, are no lighter than breeze-block walls, which partly explains the slow take-up of 3D printing technology for printing individual houses. Other structural systems therefore need to be explored if a substantial gain in mechanical performance is to be achieved. A mechanically optimized chain-linked masonry principle has been proposed by Duballet et al. (2019) on the basis of a spatial lattice. The idea is to replace a cinder block and mortar wall with a wall in which the mortar is printed in the form of a spatial lattice and separated by insulating blocks. A prototype on a representative scale has been produced (Duballet et al. 2020). Such a principle seems to be justified first and foremost by the considerable weight reduction it enables, compared with traditional construction systems (see Table 5.2).

Wall system	Thickness (cm)	Weight (kg/m^2)	U-value (W.m^{-2}.K^{-1})
Cinder blocks	40	180	0.1
Lightweight concrete	56	150	0.09
Precast wall	42	220	0.15
3D-printed truss	42	50	0.09

Table 5.2. *Comparison of the performance of different chain-linked masonry systems*

5.3.3. *Shell structures*

One strategy for avoiding bending in structural elements is to introduce curvature. This strategy is ancient and has been widely used throughout history with materials that have low tensile strength to make arches, domes, vaults and cupolas. Formwork was often used to build the structure, which was then skillfully removed. The final shape was designed not to bend, but simply to be funicular, with only compressive stresses acting on the structure. Clever builders (Persians, Romans, Byzantines, Brunelleschi, etc.) have long found constructive strategies in stone or brick to create vaults without formwork, as Auguste Choisy reports in his reference work (Choisy 1883). It is possible to draw an analogy between the constructive arrangements specific to masonry and 3D printing, as shown in the work of Carneau et al. (2020) and Motamedi et al. (2022). In 3D printing, the filament replaces the brick (Figure 5.9), and the stacking of layers replaces that of brick or cut stone beds. In addition, the materials used are frictional materials with greater compressive strength than tensile strength. It is therefore possible to transfer construction techniques originally developed for masonry vaults to 3D printing, and thus dispense with formwork. However, this requires the structure to be in equilibrium with compressive forces at every stage of its construction.

The stability of the printed structure can be assessed at both filament and structure level. Carneau proposes a classification of dispensing strategies, which are linked to the robotic system. The stability of the structure depends on the shape and the toolpath strategy adopted. When stacking filaments in 3D printing, there are three possible overhang scenarios (classified by Carneau as f^i and illustrated in Figure 5.10):

- f^0: the layer is only subjected to local shearing;
- f^1: the layer is subjected to a combination of shear and bending stresses;
- f^2: the layer is subjected to pure bending.

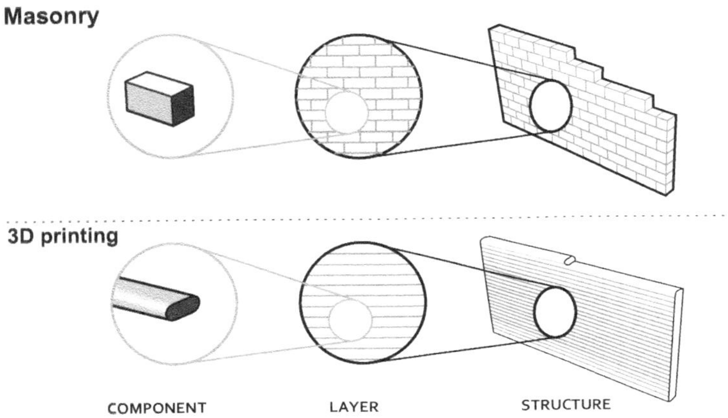

Figure 5.9. *Analogy between masonry and 3D concrete printing, according to Carneau et al. (2020)*

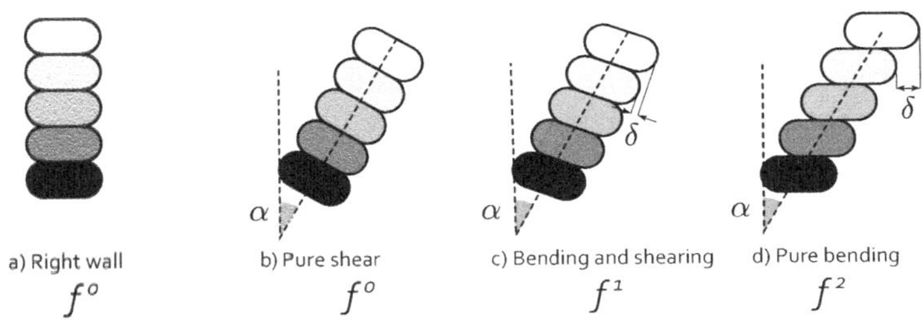

Figure 5.10. *Cord placement strategies according to Carneau et al. (2020)*

These strategies are also found in masonry structures, with ashlar masonry, Nubian vaults and corbelled domes. In addition to local filament stability, we can add the presence of intermediate supports, denoted by the superscript s^i, and layer continuity, denoted by the superscript c^i. Carneau et al. (2020) associate these strategies with a classification that is linked to robotic complexity.

Figure 5.11 shows examples of construction typologies depending on the layer deposition strategy, but also on the overall behavior of the structure: the first row corresponds to biaxial compression states, the second to arched behavior and the last row to functioning combining compression and tension. In the first column, the case of the barrel vault shows the advantage that 3D printing can have over masonry. With 3D printing, and the layer pressing strategy in particular, the cross-section of the layers can differ from a rectangular shape. Consequently, if the extrusion nozzle is always aligned with the surface, the contact area between one layer and the next is optimal and no local overhangs appear (Gosselin et al. 2016). In the second column, the 3D-printed dome shown in the table is the result of this behavior. The layers are printed orthogonally to the surface norm. The pure bending strategy in the right-hand column is difficult to apply to concrete 3D printing due to the low strength of the fresh material. For example, a layer laid horizontally but with a local overhang is very likely to over-deform.

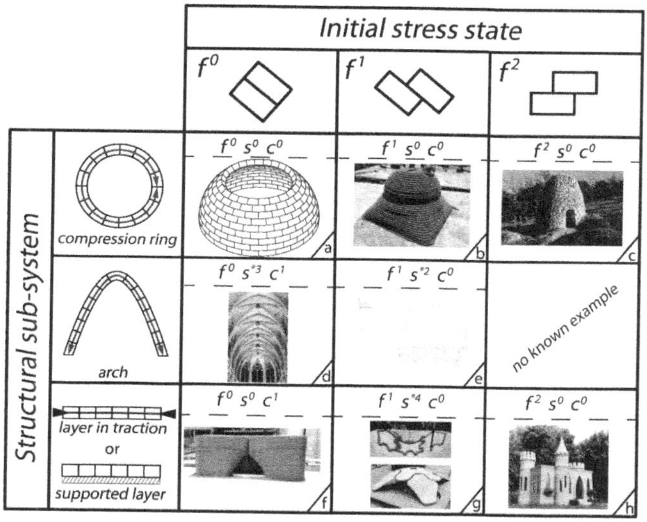

Figure 5.11. *Examples of structure types according to printing mode (Carneau et al. 2020)*

5.3.4. *Link between process and structural behavior*

The correlation between the brick's initial stress state criterion and the ability to control the geometry of the extruded layer is straightforward. These two aspects can also be aligned with material properties. A layer printed with a material with a high initial yield strength (the case of infinite brick extrusion) takes the shape of the nozzle. This is the best choice when the layer has to withstand high initial stresses, such as bending moments caused by a local overhang: bending and shearing.

On the other hand, extrusion with a very low initial yield strength (for example, with the Layer Pressing strategy) results in a layer capable of withstanding only a low shear stress when exiting the nozzle. However, it does offer the possibility of modifying its cross-sectional geometry. This relationship between material properties, initial stress state and control of layer geometry is illustrated in Figure 5.11.

5.3.5. *Toward a stereotomic approach to printing*

We have just seen that the rheological strategy employed has a direct impact on the cantilever layer deposition strategy. The analogy with masonry allows to extrapolate more complex morphological strategies than the simple superposition of a few layers. The construction of certain vaults without arches requires a special layout. For example, a barrel vault can only be stable without intermediate support if the brick bed is not arranged along the meridians (configuration $f^0 s^{*2} c^0$). Figure 5.12 shows an example of such a strategy, known as Nubian vaulting, where the brick beds (or print beads) are contained in a plane, making an angle of 24° with the vertical. It can clearly be seen that there is a variable angle between the normal to the surface (blue lines) and the plane containing the curve (orange lines), but also with the vertical direction. There is thus a torsion of the brick bed or layer. Such twisting is difficult to achieve manually and requires a six-axis robot instead of traditional systems. Here, the robot demonstrates its advantages over the human hand in terms of accuracy and speed.

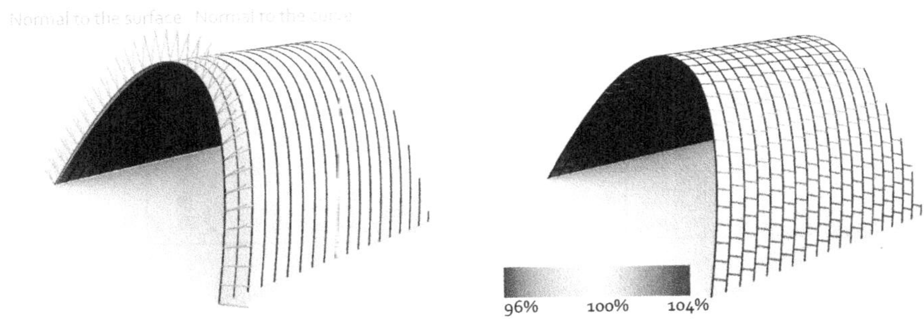

Figure 5.12. *Nubian vault. Left: inclination of bead plane (orange) and surface normal (blue). Right: relative distance between layers*

In addition, the distance between the curves drawn on the surface is variable. In practice, this means that printing along these lines results in crushing of the printed filaments. Rheological strategies based on a very significant evolution of the shear yield stress over time, typical of two-component approaches, can be exploited to achieve significant layer thickness variations, and therefore a more developed formal language. However, the crushing force due to layer thickness variation must be taken into account in the stability of the structure (Carneau et al. 2022). This approach can be compared with stone-cutting geometries that result in very fine joints, in contrast to brick structures.

It is therefore possible to study the most complex brickwork techniques and draw inspiration from them to create tool paths compatible with printing without intermediate supports. Motamedi et al. (2022) have proposed a systematic approach based on generative grammars to create printed "fixtures" whose stability is then assessed using the finite element method. This grammar is based on the basic elements of bizantine vault construction: trunks, pendentives and the Nubian apparatus of barrel vaults, which provide a better distribution of forces than naive printing paths. Figure 5.13 shows a comparison between an impression path generated by this approach and a hypothesis formulated by Eugène Viollet-le-Duc concerning the load-bearing system of Agrippa's pantheon in Rome. It appears that Viollet-le-Duc's highly complex structural language can be transcribed digitally, combining aesthetics and mechanical performance.

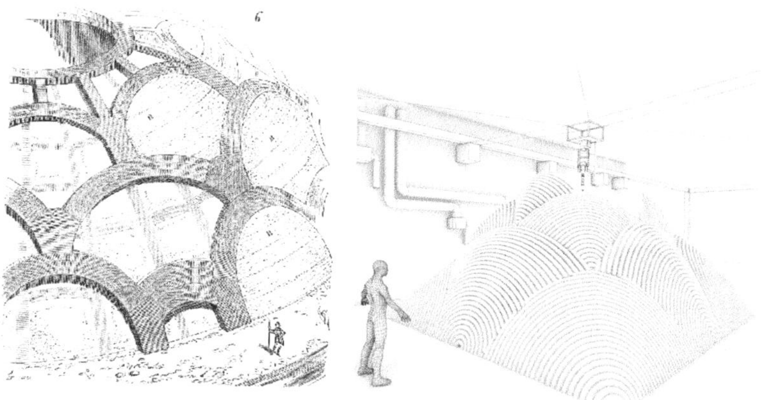

Figure 5.13. *Viollet-le-Duc's hypothesis on the structure of Agrippa's Pantheon in Rome (Viollet-le-Duc 1866) (left) and printing path generated by a grammar (Motamedi et al. 2022) (right)*

The range of possibilities offered by 3D printing of vaults therefore remains largely unexplored. Structural designers need to reappropriate a formal and technical language that has fallen into disuse. This cultural reappropriation must also take place for printing applications in reinforced concrete.

5.4. 3D printing and reinforced concrete

We have seen that 3D printing offers significant potential for formal exploration of non-reinforced masonry structures. Conventional 3D printing does not generally allow conventional reinforcement to be introduced during material deposition. However, calculation codes systematically allow for the presence of reinforcements that give the material intrinsic ductility in seismic zones. The integration of elements that provide ductility (fiber or reinforcement) is therefore a major technical challenge in the development of structural applications for 3D printing.

5.4.1. *Lost or collaborative formwork?*

The first structural applications of printed material in construction are those where it is used as formwork in which conventional reinforced concrete or fiber-reinforced concrete can be used, in compliance with an existing calculation code. There are a multitude of ways in which formwork can be used, as illustrated in Figure 5.14. The formwork can contribute to the mechanical strength of the structure, in which case it is referred to as collaborative formwork. However, the first projects carried out in Europe seem to favor an approach where the formwork is not collaborative, as printed concrete does not yet fall within an established normative framework.

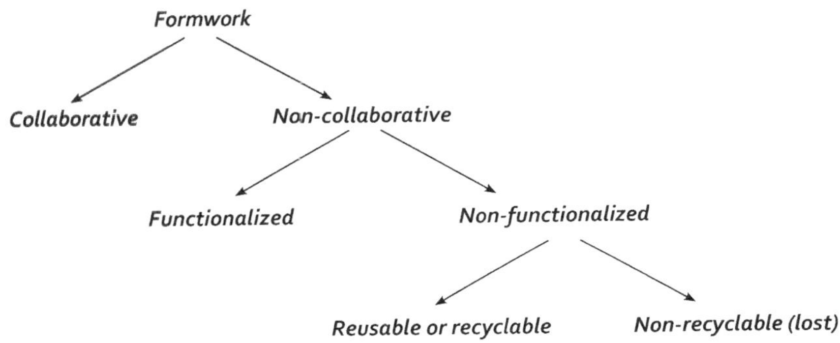

Figure 5.14. *A tree of possibilities for printed formwork*

The second level of use of formwork is to make it functional. Potential functions include:

– contribution to fire resistance;

– corrosion protection of reinforcement;

– thermal insulation;

– acoustic insulation;

– the finishing.

Finally, the last level of formwork use is its potential re-use or recycling. When formwork is not collaborative, functional, recyclable or reusable, it is referred to as lost formwork. Figure 5.15 shows an example of non-collaborating formwork. Maitenaz et al. (2024) proposed printing clay to make reusable formwork for a reinforced concrete beam optimized to Eurocode 2. The formwork can be recovered simply by crushing it. Among other examples of non-collaborating formwork, Furet et al. (2019) have proposed printing polyurethane formworks in order to create pre-walls with formwork and thermal insulation function.

Recy(clay)ble formwork: a recyclable formwork for an optimised structure

Figure 5.15. *Example of non-collaborative formwork*

Figure 5.16 is an example of lost formwork, where ultra-high-performance fiber-reinforced concrete (UHPC) is cast in printed concrete moulds. In this case, the contribution made by the printed concrete to the mechanical strength of the whole is ignored. However, the complex geometry made possible by the mould enables a reduction in material consumption to be achieved, compared with the more conventional use of fiber-reinforced concrete (Gaudillière et al. 2019).

Figure 5.16. *Fiber post in Aix-en-Provence (printing: XTreeE). Photo by Lisa Riciotti*

The issue of adhesion between the printed material and the cast-in-place material is more significant here than in confined masonry applications. In fact, the contact zone between the printed concrete and the cast concrete is potentially highly sheared or even in tension. The risk of delamination is a source of risk on two counts. First, it can cause the printed skin to detach, potentially injuring the occupants of a building: damage caused by non-structural elements must be considered by structural engineers. Second, the transfer of forces is no longer ensured after delamination and the structural operation planned by the engineer is not operational, so the mechanical strength of the assembly becomes uncertain.

Although several projects have used lost formwork, the scientific literature explaining the precise origin of potential delaminations remains tenuous. We might mention the recent work by Wang et al. (2022) and the hypotheses put forward of shrinkage deformations in the cast concrete, or of water transfer between the cast and printed concrete, which would modify the microstructure of the former.

5.4.2. *Reinforced printed concrete*

Several processes for reinforcing printed structures have been proposed in recent years. The aim of these tests is to use a technical vocabulary close to that of reinforced concrete, which is already standardized. However, there is a certain diversity in the proposals, which is reminiscent of the first concrete reinforcement systems in the 20th century, before the advent of the stirrup system developed and popularized by Hennebique (Delhumeau 1999).

The solutions explored and classified by Mechtcherine et al. (2021) include transverse screwing, printing on wire mesh following the example of ferro-cement invented and popularized by Pier Luigi Nervi, inserting reinforcement or co-extruding cable in the longitudinal direction. Kloft et al. (2020) propose a mapping of reinforcement possibilities for digitally constructed structures. Although the mapping is incomplete (long fiber reinforcement is ignored), it gives a good overview of the technological possibilities to date.

Reinforcement insertion is a technique designed to create a strengthening system identical to that commonly used in reinforced concrete construction. This similarity with current practice is an undeniable asset, and it is therefore important to understand the limits and potential of this process. The adhesion between the reinforcement and the printed material is a particular point of attention: in fact, the reinforcement may not be perfectly impregnated with concrete. The study of co-extruded cable by the Technical University of Eindhoven showed that the quality of adhesion between a cable and a printed cementitious matrix may be poorer than that of specimens cast in molds. The authors of this study attribute this poorer performance to the presence of a vacuum under the cable (Bos et al. 2020). Impregnation quality is obviously a result of the process, and there is a strong dependence on the kinematic parameters of the robot and fluid, but also on the yield stress of the deposited material. First, it is clear that if the speed of the robot and the fluid leaving the nozzle are different, the cable will tend to loosen and damage the cementitious matrix. Second, it is easy to understand that a fluid mortar will easily embed reinforcement, whereas a mortar with high yield stress will leave voids between mortar and reinforcement. Experimental study has illustrated the influence of material properties on the quality of bonding between rebars and 3D printed mortar. Numerical fluid mechanics simulations can reproduce the experimental observations and could be used to calibrate process parameters compatible with uniform reinforcement embedding (Mollah et al. 2023).

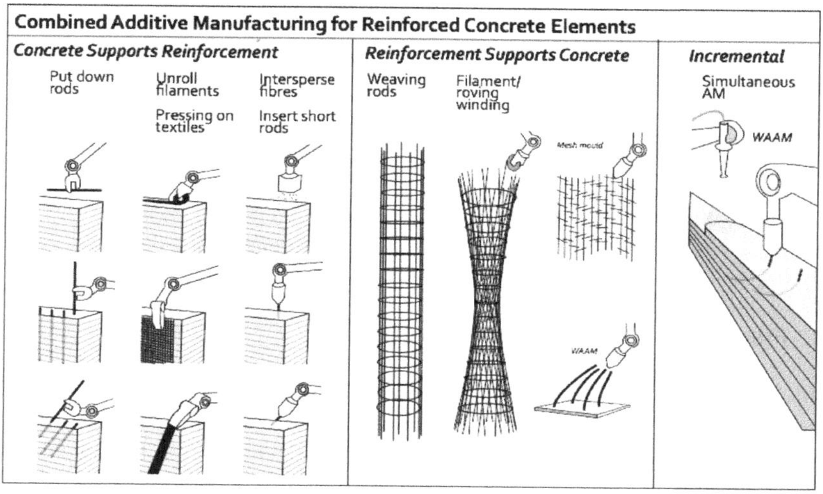

Figure 5.17. *3D printing reinforcement methods (adapted from Kloft et al. (2020))*

This observation shows that there is a contradiction between 3D printing processes, where the evolution of the yield point is a necessary condition for the part to hold during printing, and reinforced concrete processes, where the shear yield stress must be low enough to allow the mortar to flow around the reinforcement.

However, some research has succeeded in reconciling these constraints. Figure 5.18 shows a beam optimized for a point load case applied at mid-span. The beam is printed in concrete, and reinforcement is deposited according to the principle illustrated in Figure 5.17. The beam is printed on the field, resulting in a low printing height and therefore low stresses. If we combine this with the long tool path between each layer, we can deduce that the constructibility requirements of the printed material are lower than for many of the applications considered so far (e.g. walls several meters high). In the same way, we could imagine printing a ribbed slab, using a material with a low yield stress and automated reinforcement placement with good impregnation.

Figure 5.18. *Printed and reinforced beam (SDU.Create, 3D printing: Hyperion Robotics) (Breseghello et al. 2023)*

5.4.3. *Fibering printed concrete*

The integration of reinforcement in printed structures can be problematic, as their implementation is subject to practical constraints. In fact, a minimum cover distance[2] must be respected to ensure the transfer of forces from the reinforcement to the concrete, and to guarantee the durability of the structure. Typical cover values for a building with little exposure are around 3 cm, and consequently printed wall thicknesses of at least 6 cm. However, a highly efficient structure can be thinner than this. By way of example, the thickness of the concrete skins on the shell of the CNIT[3] is 6 cm. For these reasons, fibering is a serious avenue of research for expanding the potential application of 3D printing to structural elements (see Chapter 6).

5.5. Prestressing

5.5.1. *Principle*

The principle of prestressing concrete was perfected and popularized by Eugène Freyssinet in the mid-20th century. It consists of a simple observation: concrete has very good compressive strength and poor tensile strength. Compressive failure of

2 The minimum distance between the reinforcement and the free edge of the concrete.

3 This shell was designed by Nicolas Esquillan and built in 1958. With a span of 218 m, it is the longest span for a concrete shell to date.

concrete is quasi-ductile: concrete could therefore be used without reinforcement, or with very little reinforcement if it were subjected only to compression. Prestressing involves placing a structure under compression using a high-strength wire rope. In this way, a state of self-stress is created, superimposed on the stresses resulting from external loads. Cable tensioning can be accompanied by the injection of a cementitious grout, the main advantage of which is to protect the steel from corrosion. This is known as *bonded prestressing*.

5.5.2. *Application*

Several examples of small-scale gateways have been built along similar lines (Salet et al. 2018; Vantyghem et al. 2020). First, voussoirs are printed, then passive steels used to ensure shear strength are added. In the case of Salet et al. (2018), a cable is co-extruded with the mortar, while a manual step is required for the footbridge shown in Figure 5.19. The voussoirs are then pre-positioned and post-tensioned by jacking.

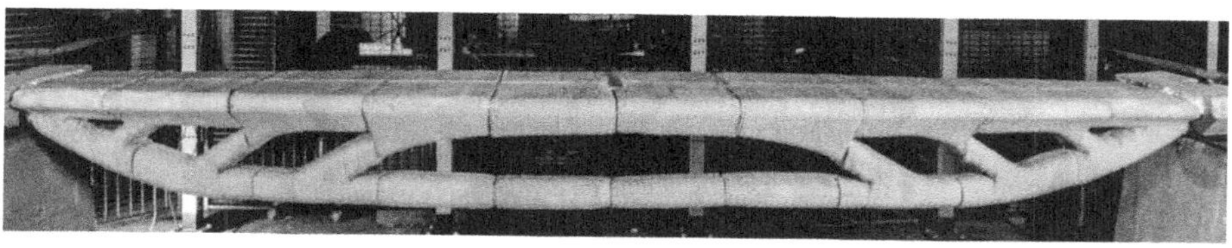

Figure 5.19. *Post-tensioned bridge with 3D printed voussoirs, adapted from Kloft et al. (2024)*

Post-tensioning provides a means of connecting prefabricated elements, and thus offers an attractive avenue of application for optimized printed elements assembled by post-tensioning. The principle of post-tensioning can be extended to any flexural system, including slab or vault systems. The design possibilities for these highly hyperstatic systems are far greater, since an infinite number of self-tensioning states can be generated, offering significant potential for material optimization compared with simple beams. Numerically, these self-stressing states are found by calculating the kernel of the equilibrium matrix, which can be done with linear algebra libraries (Pellegrino and Calladine 1986).

5.5.3. *Challenges*

The examples given show that the use of prestressing in printed concrete is a particularly promising approach. It is close to the norm, compatible with all printing systems and rheological strategies, and highly effective. However, prestressing remains a highly technical process that requires great dimensional precision and control of the surface conditions of the contact zones. Knowledge of the creep of the structure is also necessary to control the loss of prestressing over time. In fact, it was a detailed understanding of this phenomenon that enabled Freyssinet to successfully apply prestressing. As 3D printing is an emerging technology, we are not aware of any studies on the creep of printable mortars.

The use of prestressing combined with the formal richness offered by 3D printing also opens up a field of morphological exploration that is still too unexploited. Here again, we can turn to history, and more specifically to the structural artist, Eladio Dieste. Dieste's work, now a UNESCO World Heritage Site, is well-known for its projects of masonry structures prestressed by a clever system of loops used as mechanical amplifiers. The combination of brick, reinforced concrete and prestressing has enabled him to build slender vaults of great elegance. Prestressing printed vaults without formwork is a very promising avenue of development, which could combine form strength and prestressing.

Figure 5.20. *Eladio Dieste's slim hulls: prestressing system (left) and hull. Image: Pedreschi (2006)*

5.6. Conclusion

Printed concrete is still mainly used for non-structural applications. The use of printed concrete as a confined masonry element has made it possible to gain acceptance on building sites, but work on the shape and the integration of reinforcements are the two avenues that offer the prospect of significant performance gains for these structures.

Using 3D printing to build vaults requires not only a study of mechanics, but also of the art of stereotomy. If a gantry-mounted printer is sufficient to print a vertical wall, the use of six-axis robots seems almost inevitable to print self-stabilizing vaults without temporary support. The formal language of stereotomy and masonry can be reinterpreted to generate tool paths that are far more complex and efficient than the 2.5D paths found in most printed objects.

A profusion of reinforcement ideas has emerged over the last few years, with none of them gaining the upper hand. We can imagine that only a handful of reinforcement technologies will emerge as winners from the economic selection process that will take place over the next few years, as was the case with the Hennebique system. Only tangible projects and large-scale deployment of these processes will allow them to emerge.

5.6.1. *Structural optimization and additive manufacturing*

The optimization of structures, and in particular topological optimization, seems an obvious avenue of research, so much so that it is highlighted in numerous publications. Indeed, the formal freedom offered by 3D printing makes it possible to imagine significant material savings. Studies on hollowed-out beams by Maitenaz et al. (2024) show that 30% on bent systems are conceivable. Digital manufacturing seems to be bringing us closer to the materialization of optimal structures.

However, it is important to use this kind of tool with the necessary distance and humility. Optimization is the numerical resolution of a mathematical problem based on the choice of a physical model. This model does not necessarily represent constructive and mechanical reality, and it is up to the engineer to use the model that is as faithful as possible and as easy to explain as possible. We have shown, for example, that plate models can describe printed walls, taking care to calculate equivalent homogeneous properties.

The choice of the objective function to be minimized is also crucial. The overwhelming majority of topological optimization consists of solving a flexibility minimization problem, which is better suited to ULS for elastic materials and ignores SLS issues. Recent developments in topological optimization at ultimate limit states, based on fracture calculations, seem to be particularly promising (Mourad et al. 2021, 2022). Finally, the designer must bear in mind the quote from Archie R. Dykes and accept the uncertain nature of structural design. Instead of deterministic optimization, the future of optimization may lie in robust optimization, which takes into account load or material uncertainties.

5.6.2. *The structural language of 3D printing*

Finally, while the engineer's primary role is to ensure structures can withstand external loads without failure, and save resources, they also act as a cultural actor, using a technique to give shape to buildings or infrastructures. In so doing, they inscribe the structure in a constructive history, whether in continuity or rupture. A new technology leads to the creation of a new structural language. We have seen that Eladio Dieste's shells offer a technical synthesis of all the strategies discussed in this chapter:

masonry, reinforcement and prestressing, using mid-20th-century tools. Large-scale experimentation and risk-taking, made possible by advanced digital tools, will enable us to carry out this work.

5.7. References

Asprone, D., Menna, C., Bos, F.P., Salet, T.A., Mata-Falcón, J., Kaufmann, W. (2018). Rethinking reinforcement for digital fabrication with concrete. *Cement and Concrete Research*, 112, 111–121.

Baverel, O., Douthe, C., Mesnil, R. (2022). Conception des structures architecturales dans un contexte de ressources limitées. In *Transitions : les nouvelles annales des ponts et chaussées - bâtiments et construction en transition*, Vandamme, M. (ed.). Presses des Ponts, Paris.

Bažant, Z.P. (1999). Size effect on structural strength: A review. *Archive of Applied Mechanics*, 69, 703–725.

Bos, F., Dezaire, S., Ahmed, Z., Hoekstra, A., Salet, T. (2020). Bond of reinforcement cable in 3D printed concrete. In *Second RILEM International Conference on Concrete and Digital Fabrication*, Bos, F.P., Lucas, S.S., Wolfs, R.J., Salet, T.A. (eds). Springer, Cham.

Breseghello, L., Hajikarimian, H., Jørgensen, H.B., Naboni, R. (2023). 3DLightBeam+. Design, simulation, and testing of carbon-efficient reinforced 3D concrete printed beams. *Engineering Structures*, 292, 116511. doi: 10.1016/j.engstruct.2023.116511.

Carneau, P., Mesnil, R., Roussel, N., Baverel, O. (2020). Additive manufacturing of cantilever – From masonry to concrete 3D printing. *Automation in Construction*, 116, 103184 [Online]. Available at: https://www.sciencedirect.com/science/article/pii/S0926580519308568.

Carneau, P., Mesnil, R., Baverel, O., Roussel, N. (2022). Layer pressing in concrete extrusion-based 3D-printing: Experiments and analysis. *Cement and Concrete Research*, 155, 106741 [Online]. Available at: https://www.sciencedirect.com/science/article/pii/S0008884622000321.

Caron, J.-F., Baverel, O., Douthe, C., Mesnil, R., Gobin, T. (2018). Proposals to make complex structures affordable. In *IABSE Symposium, Nantes 2018: Tomorrow's Megastructures*. AFGC.

Caron, J.-F., Demont, L., Ducoulombier, N., Mesnil, R. (2021). 3D printing of mortar with continuous fibres: Principle, properties and potential for application. *Automation in Construction*, 129, 103806 [Online]. Available at: https://www.sciencedirect.com/science/article/pii/S0926580521002570.

Choisy, A. (1883). *L'art de bâtir chez les Byzantins*. Société anonyme de publications périodiques, Paris.

Delhumeau, G. (1999). *L'invention du béton armé : Hennebique, 1890-1914*. Editions Norma, Paris.

Duballet, R., Baverel, O., Dirrenberger, J. (2019). Space truss masonry walls with robotic mortar extrusion. *Structures*, 18, 41–47 [Online]. Available at: https://www.sciencedirect.com/science/article/pii/S2352012418301309.

Duballet, R., Mesnil, R., Ducoulombier, N., Carneau, P., Demont, L., Motamedi, M., Baverel, O., Caron, J.-F., Dirrenberger, J. (2020). Free deposition printing for space truss structures. In *Second RILEM International Conference on Concrete and Digital Fabrication: Digital Concrete 2020*. Springer, Heidelberg.

Furet, B., Poullain, P., Garnier, S. (2019). 3D printing for construction based on a complex wall of polymer-foam and concrete. *Additive Manufacturing*, 28, 58–64. doi: 10.1016/j.addma.2019.04.002

Gaudillière, N., Duballet, R., Bouyssou, C., Mallet, A., Roux, P., Zakeri, M., Dirrenberger, J. (2019). Building applications using lost formworks obtained through large-scale additive manufacturing of ultra-high-performance concrete. In *3D Concrete Printing Technology*, Sanjayan, J.G., Nazari, A., Nematollahi, B. (eds). Butterworth-Heinemann, Oxford [Online]. Available at: https://www.sciencedirect.com/science/article/pii/B9780128154816000038.

Gosselin, C., Duballet, R., Roux, P., Gaudillière, N., Dirrenberger, J., Morel, P. (2016). Large-scale 3D printing of ultra-high performance concrete: A new processing route for architects and builders. *Materials & Design*, 100, 102–109 [Online]. Available at: https://www.sciencedirect.com/science/article/pii/S0264127516303811.

van den Heever, M., du Plessis, A., Kruger, J., van Zijl, G. (2022). Evaluating the effects of porosity on the mechanical properties of extrusion-based 3D printed concrete. *Cement and Concrete Research*, 153, 106695 [Online]. Available at: https://www.sciencedirect.com/science/article/pii/S0008884621003446.

Kloft, H., Empelmann, M., Hack, N., Herrmann, E., Lowke, D. (2020). Reinforcement strategies for 3D-concrete-printing. *Civil Engineering Design*, 2(4), 131–139.

Kloft, H., Sawicki, B., Bos, F., Dörrie, R., Freund, N., Gantner, S., Gebhard, L., Hack, N., Ivaniuk, E., Kruger, J. et al. (2024). Interaction of reinforcement, process, and form in digital fabrication with concrete. *Cement and Concrete Research*, 186, 107640. doi: 10.1016/j.cemconres.2024.107640.

Kruger, J. and van der Westhuizen, J.-P. (2023). Investigating the Poisson ratio of 3D printed concrete. *Applied Sciences*, 13(5) [Online]. Available at: https://www.mdpi.com/2076-3417/13/5/3225.

Maitenaz, S., Mesnil, R., Feraille, A., Caron, J.-F. (2024). Materialising structural optimisation of reinforced concrete beams through digital fabrication. *Structures*, 59, 105644. doi: 10.1016/j.istruc.2023.105644.

Marchment, T. and Sanjayan, J. (2020). Mesh reinforcing method for 3D concrete printing. *Automation in Construction*, 109, 102992.

Mechtcherine, V., Buswell, R., Kloft, H., Bos, F.P., Hack, N., Wolfs, R., Sanjayan, J., Nematollahi, B., Ivaniuk, E., Neef, T. (2021). Integrating reinforcement in digital fabrication with concrete: A review and classification framework. *Cement and Concrete Composites*, 119, 103964 [Online]. Available at: https://www.sciencedirect.com/science/article/pii/S0958946521000342.

Mechtcherine, V., van Tittelboom, K., Kazemian, A., Kreiger, E., Nematollahi, B., Nerella, V.N., Santhanam, M., de Schutter, G., Van Zijl, G., Lowke, D. et al. (2022). A roadmap for quality control of hardening and hardened printed concrete. *Cement and Concrete Research*, 157, 106800 [Online]. Available at: https://www.sciencedirect.com/science/article/pii/S0008884622000916.

Mesnil, R., Poussard, V., Sab, K., Caron, J.-F. (2023). On the geometrical origin of the anisotropy in extrusion-based 3D printed structures. *Engineering Structures*, 275, 115082 [Online]. Available at: https://www.sciencedirect.com/science/article/pii/S0141029622011580.

Mollah, M.T., Comminal, R., Leal da Silva, W.R., Šeta, B., Spangenberg, J. (2023). Computational fluid dynamics modelling and experimental analysis of reinforcement bar integration in 3D concrete printing. *Cement and Concrete Research*, 173, 107263 [Online]. Available at: https://www.sciencedirect.com/science/article/pii/S0008884623001771.

Motamedi, M., Mesnil, R., Oval, R., Baverel, O. (2022). Scaffold-free 3D printing of shells: Introduction to patching grammar. *Automation in Construction*, 139, 104306 [Online]. Available at: https://www.sciencedirect.com/science/article/pii/S0926580522001790.

Mourad, L., Bleyer, J., Mesnil, R., Nseir, J., Sab, K., Raphael, W. (2021). Topology optimization of load-bearing capacity. *Structural and Multidisciplinary Optimization*, 64(3), 1367–1383.

Mourad, L., Bleyer, J., Mesnil, R., Nseir, J., Sab, K., Raphael, W. (2022). Multi-material topology optimization of structural load-bearing capacity using limit analysis. Working Paper, HAL [Online]. Available at: https://enpc.hal.science/hal-03528107/.

Pedreschi, R. (2006). The structural behaviour and design of free-standing barrel vaults of Eladio Dieste. In *Proceedings 2nd International Congress on Construction History*. Short Run Press, Exeter.

Pellegrino, S. and Calladine, C.R. (1986). Matrix analysis of statically and kinematically indeterminate frameworks. *International Journal of Solids and Structures*, 22(4), 409–428. doi: 10.1016/0020-7683(86)90014-4.

Salet, T.A., Ahmed, Z.Y., Bos, F.P., Laagland, H.L. (2018). Design of a 3D printed concrete bridge by testing. *Virtual and Physical Prototyping*, 13(3), 222–236.

Vantyghem, G., De Corte, W., Shakour, E., Amir, O. (2020). 3D printing of a post-tensioned concrete girder designed by topology optimization. *Automation in Construction*, 112, 103084 [Online]. Available at: https://www.sciencedirect.com/science/article/pii/S0926580519305096.

Viollet-le-Duc, E.-E. (1866). *Dictionnaire raisonné de l'architecture française du XIe au XVIe siècle*. Morel, Paris.

Wang, L., Yang, Y., Yao, L., Ma, G. (2022). Interfacial bonding properties of 3D printed permanent formwork with the post-casted concrete. *Cement and Concrete Composites*, 128, 104457 [Online]. Available at: https://www.sciencedirect.com/science/article/pii/S095894652200052X.

6

Reinforcement of Printed Structures

Jean-François CARON[1], Nicolas DUCOULOMBIER[2] and Léo DEMONT[1]

[1] *Laboratoire Navier, École des Ponts, Marne-la-Vallée, France*

[2] *XtreeE, Rungis, France*

With a tensile strength of just a few MPa and brittle behavior, cementitious mortars cannot legally be used unreinforced under tensile loading configurations. This is true today for cast structures, and will be so in the future for printed structures. Reinforcing printed mortar is therefore a highly strategic issue for the development of the technology. In this chapter, we first review the history and technologies of cement and concrete reinforcement in order to understand the mechanisms involved. In the second part, we take stock of initiatives concerning 3D printing in particular, and detail an original technology for reinforcing mortar with long, continuous fibers.

6.1. Introduction, a few reminders about the reinforcement of cementitious materials

Cementitious materials have a much lower tensile strength than compressive materials, as well as a high degree of brittleness. This behavior is due to the presence of natural micro-cracks, which form during setting and propagate under load to the point of failure, since the toughness of these matrices is low. A micro- or macro-crack will have a greater impact on the material's behavior, and the more it is oriented perpendicularly to the tensile forces. The solution, therefore, is to "sew"

3D Concrete Printing,
coordinated by Arnaud PERROT and Yohan JACQUET.

these cracks with a suitable reinforcement, allowing the material to be discharged locally in favor of the reinforcement system.

Two types of reinforcement exist for concrete and mortar: those that we shall refer to as structural, such as reinforcement and cables, and those that are more on the scale of the material, such as fiber reinforcements. We will come back to these two families in the following section so as to fully understand the mechanics of these mixes, and to identify the issues and problems that will remain the same for reinforced printed mortars.

6.1.1. *Structural reinforcement of cementitious materials*

The oldest and best-known technique, due to its widespread use, is reinforced concrete, that is, concrete reinforced by the addition of metal reinforcement to take up tensile stresses throughout the structure. The idea is attributed to the French engineer Joseph Lambot, who in 1848 first came up with the idea of reinforcing the cement of his agricultural ferries, and then that of a small boat, with a steel lattice. He patented the concept in 1855, which became known as "Ferrocement" (Naaman 2000). Numerous developments followed, notably in the building and civil engineering sectors, with no fewer than 262 patents registered between 1886 and 1906, leading to the first regulations in France that same year. There were two opposing approaches. The first proposed post-and-beam construction systems, with the progressive use of steel sections with larger cross-sections, known as "concrete reinforcing bars" or steel rebars. This approach, initially promoted by François Hennebique, is the one that finally took hold and still accounts for a significant proportion of today's constructions. The second approach, on the other hand, uses fine-diameter reinforcements forming a steel network to ensure a strong adhesion between the reinforcement and the matrix. This system, promoted by Cottencin, was taken up by Italian architectural engineer Pier Luigi Nervi in the middle of the last century, who went on to develop the concept of fine steel mesh. "Ferrocement" then became a technique that enabled architectural feats such as the hangar at the Orbetello airfield (Italy) in 1940 or Rome's Palazzetto dello Sport in 1957 (Figure 6.1) to be achieved, with its staggering efficiency and inspiring aesthetics. However, its delicate implementation requires special expertise, and its high cost condemns it to niche applications. We will come back to this in the conclusion to this chapter, as the underlying concepts were so visionary and adapted to today's means.

Another method for reinforcing concrete structures, developed and patented by Eugene Freyssinet in 1928, is prestressing. By superimposing a sufficiently high local compressive stress on the external load, this method allows the material element to work only in compression, thus avoiding (if this prestressing does not relax over time) the problem of its poor tensile strength. This method also has the advantage of enabling dry assembly of prefabricated concrete elements, widely used in the manufacturing of bridges and overhanging structures. These two reinforcement methods are sometimes used simultaneously, as part of structural systems that compensate for material weakness.

Figure 6.1. *Palazzetto dello Sport, Rome, 1957 (GiulioCesare, archidiap.com/opera/palazzetto-dello-sport/, original dimensions, licensed under CC BY-SA 4.0)*

6.1.2. *Fibering of cementitious materials*

There are also reinforcement methods that improve mechanical behavior at the level of the material itself. In the case of concrete, this is the role of aggregates, which increase the compressive strength and ductility of mortar by providing dissipation mechanisms. Similarly, the addition of a small quantity of fibers with high tensile strength improves tensile behavior, notably by enabling the sewing of micro- or macro-cracks. Two sub-categories can be distinguished depending on whether these composites are made from short, generally randomly oriented fibers, or from continuous fibers aligned in one or more given directions. If we are particularly interested in tensile mechanical properties, the addition of continuous fibers provides a greater improvement in the mechanical behavior of the composite in the fiber direction(s) compared to the addition of the same volume of short fibers, whose orientation is often random. However, in the first case, the gain in strength is maximum only in the direction of the continuous reinforcements, which is referred to as anisotropic behavior. In the second case, the tensile behavior of the representative elemental volume (REV) remains virtually isotropic, since it is improved in all directions.

6.1.2.1. *Reinforcing cementitious materials with short fibers*

Short-fiber reinforcement is generally achieved by introducing the fibers directly into the mixer during matrix production. However, their addition causes a significant drop in the fluidity of the mix (Martinie 2010). The volume fraction introduced is therefore limited (a few percent by volume), as is the reinforcement obtained, and the effect on rheology is highly dependent on fiber slenderness (length/diameter).

The aim of adding these fibers is to slow down or even prevent crack propagation in the cementitious matrix and improve the ductility of the whole by avoiding sudden, localized failure. They must therefore be sufficiently anchored in the cementitious matrix to be able to take up forces on either side of the crack to limit its opening, but also be able to loosen before fracture to provide ductility by dissipating energy at the fiber/matrix interfaces.

The type of effect a fiber has on overall behavior is closely linked to its diameter (Rossi 2001; Betterman and Ouyang 1995; Li and Leung 1992; Rossi 1998; Brandt 2008), which is generally correlated with its length. Sufficiently fine fibers may have an effect on microcracking and delay localization, but their short length limits their action to small crack openings, that is, before localization. This results in low tensile ductility. However, larger diameter fibers act once localization has been activated. They therefore do not increase strength, but do provide a certain ductility (see Figure 6.2).

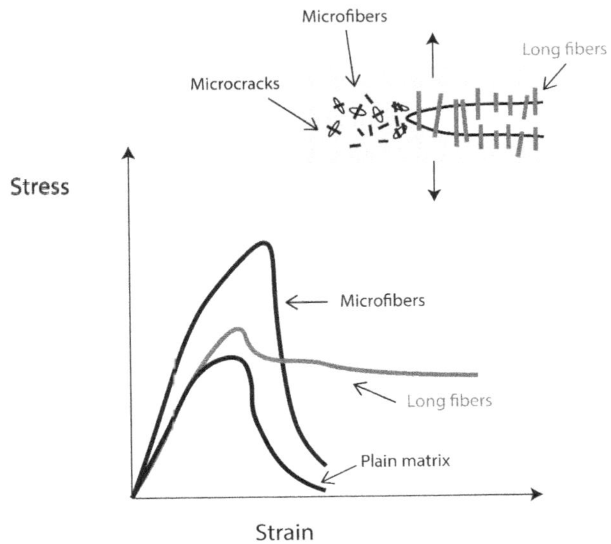

Figure 6.2. *Effect of fibers according to diameter and size (adapted from Betterman and Ouyang 1995). For a color version of this figure, see www.iste.co.uk/perrot/3dconcrete.zip*

Numerous short fibers have been developed for the reinforcement of cementitious materials (Brandt 2008; Hamidi and Aslani 2019).

– metal fibers of various sizes and geometries;

– mineral fibers, particularly AR (alkali-resistant) glass fibers, that is, those resistant to the alkalinity of the cement matrix, and more recently basalt fibers;

– synthetic fibers made from various materials, including polypropylene (PP), polyethylene (PE), nylon and polyvinyl alcohol (PVA);

– carbon fibers;

– natural fibers (e.g. hemp and miscanthus).

Metal fibers are still the most widely used reinforcement for cementitious materials, and make up the family of metal fiber concretes. To limit their influence on rheology, the slenderness value is often around 60 (Rossi 1998). They generally have a relatively weak bond with the cementitious matrix. As a result, numerous geometries have been developed to optimize their anchorage: twisted, with hooks, etc. (Brandt 2008). The functioning of metal fibers with complex geometries is described in Chanvillard (1992). As far as mechanical behavior is concerned, fiber concretes made from a single type of metal fiber always lead to the achievement of moderate reinforcement characterized mainly by a reduction in brittleness, illustrated on the force-deformation curve in Figure 6.3(a).

There are two strategies for increasing the performance of fiber-reinforced cementitious composites (high-performance fiber-reinforced cementitious composites (HPFRCC)). These involve either increasing the quantity of fiber by modifying the manufacturing process (which is difficult), or taking advantage of the synergy obtained by using fibers of different sizes and diameters (generally three types of fiber), which results in a significantly higher fiber content while maintaining good workability (Rossi 2001, 2013). The behavior is then characterized by the presence of an additional phase of positive strain-hardening behavior, even in direct tension, between the elastic and softening behavior, illustrated in Figure 6.3(b). The parameters σ_{cc} and σ_{pc} are then generally defined, corresponding to the value of the stress associated with the appearance of the first macrocrack and the maximum stress, representing the ultimate tensile strength f_t.

The range of behaviors obtained depends on the type and quantity of fiber used. In the case of metal fibers (Wille et al. 2014), yield strength and maximum direct tensile strength can reach 15 and 20 MPa, respectively, for strains of 0.03% and 0.5%, respectively.

Other types of fiber, notably synthetic fibers, have been introduced more recently. Softer and more slender, they are generally centimeter-long (Brandt 2008) and have a wide range of mechanical properties. This has given rise to a new family of cementitious materials known as engineered cementitious composites (ECC), offering a wide range of behavioral characteristics. For example, in direct tension (Li et al. 2020), strain-hardening cementitious composites (SHCC), composed of a fine cementitious matrix and a limited fraction of synthetic fibers (Li 2003), have a tensile strength of less than 5 MPa, but a strain-hardening behavior up to a very high strain-to-failure value of the order of 5% (Li 2003; Jun and Mechtcherine 2010).

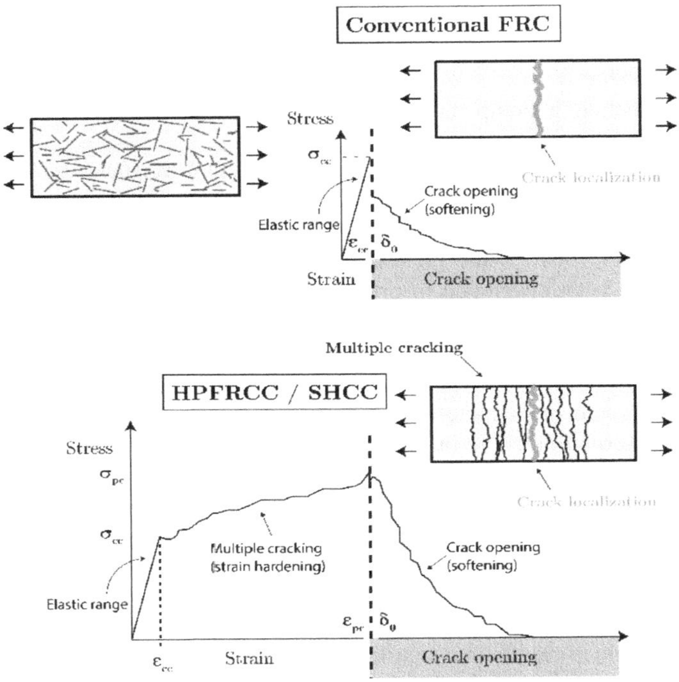

Figure 6.3. *Diagram and characteristic behavior curve (adapted from Betterman and Ouyang 1995). For a color version of this figure, see www.iste.co.uk/perrot/3dconcrete.zip*

6.1.2.2. *Continuous reinforcement of cementitious materials*

Even if we exclude from this description metal reinforcements, as described in the introduction (irons and grids), there are continuous reinforcement solutions using long fibers or threads. Mainly used to make prefabricated structural elements or to repair old structures, they have taken advantage of the recent development of low-cost mineral and organic fibers, enabling the production of high-weight rovings, textiles and nets with excellent mechanical properties and insensitivity to corrosion. There are several possible arrangements for the fibers in the roving. They can be straight, known as laminated roving, or twisted. Their particular mechanical behavior has been studied in detail by Chudoba et al. (2006). We could also mention fabric reinforcement made from the same threads, which differs from netting in its woven structure (Peled et al. 1994), but whose implementation principle is similar. These materials are sometimes called textile reinforced concrete (TRC) or textile reinforced mortar (TRM), and were first proposed for slender structures containing very fine elements (Hegger and Voss 2008; Hegger et al. 2018), which could then be pultruded (Peled and Mobasher 2006), and to reinforce aging structures (Brückner et al. 2006; Mechtcherine 2013; Kouris and Triantafillou 2018). It should be noted that these cementitious composites, like ferrocement, are made from a cementitious matrix of low particle size (absence of centimetric aggregates), often including silica fume and mineral fillers.

The threads are made from mineral or synthetic multi-fiber rovings, generally AR glass fibers. The individual fibers are of the order of 10 μm in size, and the threads contain a few thousand filaments (Hegger et al. 2006).

The simple tensile behavior of laminated composites presents three phases:

1) a linear elastic phase with no visible cracks until the first crack appears;

2) a phase of strain-hardening behavior marked by multi-cracking, the spacing of which is highly dependent on the reinforcement/matrix interface;

3) a strain-hardening phase, corresponding to the opening of existing cracks whose stiffness is slightly lower than that of the composite fabric.

Naturally, reinforcement is less effective when loading is not oriented in the main thread directions, and the quality of impregnation of the fibers by the cementitious matrix plays an essential role in the behavior of TRCs, since it dictates the transfer of load from the matrix to the reinforcement.

Tensile mechanical behavior in these main directions is in fact very similar to that of unidirectional brittle matrix composites, especially unidirectional ceramic matrix composites (CMCs) (Curtin et al. 1998; Xia and Curtin 2000; Castelier et al. 2010; Chateau et al. 2015). However, the latter benefit from far more sophisticated and costly processes, and differ in terms of much higher fiber content, and far superior impregnation qualities (vacuum and/or gas phase). Here, the particular structure of multi-fiber yarns and the imperfect impregnation of the cementitious matrix through the network of fibers that make up the yarn limit the utilization rate of the reinforcement and lead to damage to its mechanical properties during loading (Häußler-Combe and Hartig 2007). This is referred to as reinforcement efficiency, defined as the ratio of the tensile strength of the composite to the bare strength of the reinforcement (Hegger et al. 2006).

After these reminders concerning the reinforcement of cast mortars, we take stock of initiatives concerning the reinforcement of printed materials.

6.2. Reinforcement methods for additively manufactured cementitious materials and structures

Changing the process from pouring/pumping to printing/extrusion does not totally change the issues or the solutions. The main part of what has been described above will be found in the proposals for the reinforcement of printed structures, inserts, prestressing, short or long fibers, etc. However, the new process fundamentally proposes a different way of designing. Instead of a volumetric vision of the objects to be built, induced by the formwork and casting tools, it transforms them into a surface vision induced by the laying process. It thus brings new rules and imperatives into play, opening the way to new constructive systems and, consequently, to new modes of reinforcement. For example, it allows us to rethink the optimization of the shape according to a given load, just as the ancients did naturally. Nubian vaults were built without formwork, as were Freyssinet's prestressed arches and Nervi's funicular vaults, all of which benefited from cheap, abundant skilled labor, and thus compensated for the high cost of materials. The material is becoming expensive again, at least in environmental terms, and today if labor is scarce and expensive, digital technology and robotics can make up for it. This means that structures can be reexamined in terms of their shape, and pure compression, for example, can be sought without the need for direct reinforcement. In Carneau et al. (2020, 2019), printed arch and vault systems are proposed. Additive manufacturing of cementitious materials is seen as a new approach to masonry with continuous brick beds, something Pegna (1997) had already noted in 1997. There is no need for material reinforcement here, but what about the strength of the joint between layers. Experience has shown that a suitable procedure can avoid so-called cold-jointing, that is, weak joints, and even achieve a strength equal to that of the mortar at this singularity. Many studies have been carried out on this subject, however, and (Keita et al. 2019) show that the difference in strength between the deposited material and the transition zone is in fact mainly a function of the drying of the free surface before the new layer is deposited. There is also a maximum overlap time, t_r, to avoid the appearance of a joint of lower strength. Khoshnevis (2004) has proposed the use of staples to improve the strength of the interface between two successive layers, and a number of studies have also focused on this subject.

Alongside these reflections linking form, material and structure, and in order to avoid tensions, the economic deployment of the technology involves more conventional applications – quasi-vertical walls, horizontal beams and slabs – for which reinforcement of the printed material is of course necessary. Numerous approaches have been proposed in the literature, although there is as yet no reference solution. These proposals can be classified according to a criterion proposed in Asprone et al. (2018b), which groups reinforcement methods according to the time of application. For some of them, reinforcement is applied a posteriori, inspired by traditional reinforcement methods. For others, it is integrated directly into the initial batch of material. For the latter, reinforcement is carried out simultaneously with in-line printing, in the immediate vicinity of the printing nozzle. We now review the techniques available to date.

6.2.1. *A posteriori reinforcements*

Here, the structure is reinforced a posteriori, that is, once it has been printed. The classic reinforced or prestressed concrete techniques mentioned above are used. It is possible, for example, to produce structures in printed cementitious materials, incorporating recesses for inserting reinforcing bars or prestressing tendons secured by injection of a grout as proposed in Figure 6.4(c). The advantage of this type of process is that it can be rapidly deployed on construction sites, since existing dimensioning methods integrated into the regulatory framework are directly applicable.

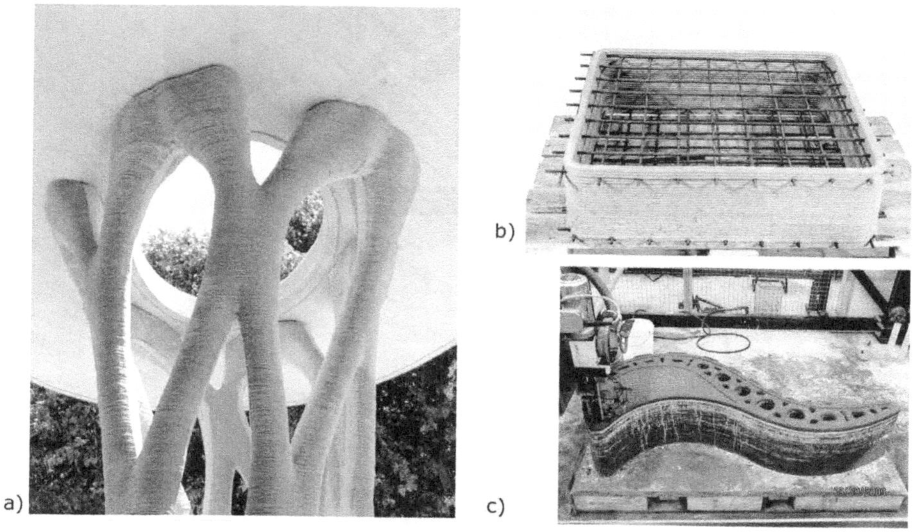

Figure 6.4. *A posteriori reinforcement: (a) column with printed formwork and BFUP XtreeE (Duballet 2018); (b) printed formwork for concrete casting reinforced with traditional reinforcement cages (photo courtesy of Spie-Batignolles); (c) printing with pre-stressed cable reservations © Richard Buswell. For a color version of this figure, see www.iste.co.uk/perrot/3dconcrete.zip*

Another way of using 3D printing to create complex regulatory structures is to print the formwork into which the concrete will be cast and reinforced using traditional methods. This is what the Chinese company Winsun or the French one Spie-Batignolles (see Figure 6.4(b)) proposes, for example, by positioning reinforcement in a printed mold. Another example is XtreeE, which prefers to use ultra-high-performance fiber-reinforced concrete, which is simpler to process, to make the column illustrated in Figure 6.4(a). In Maitenaz et al. (2021, 2024), another example concerns a methodology for optimizing a Eurocode-compliant beam using such printed formwork. Potential concrete savings are significant: thanks to a parametric study, they have been estimated at 48% on average for residential building beams. A prototype is shown in Figure 6.5, where the formwork and robotic configuration are also visible and a scale. One realisation by ISC Vinci and Chantiers Modernes construction is shown.

Figure 6.5. *Prototype of an optimized reinforced concrete beam with printed formwork (Maitenaz et al. 2024) and an application by Chantiers Modernes Construction in Drancy, France. For a color version of this figure, see www.iste.co.uk/perrot/3dconcrete.zip*

A major drawback is that today's regulations do not allow formwork cement to be taken into account in dimensioning, as the level of collaboration between hardened printed cement and freshly cast cement is unknown. So, even if this method allows for complex shapes, in some cases it can be handicapped by the additional cost, including environmental costs.

Other more recent proposals involve reinforcing printed structures with external reinforcements. In Asprone et al. (2018a), a beam is proposed in which the compressive parts are produced by robotized extrusion in cementitious material, while the tensile forces are taken up by an external metal skeleton. In Salet et al. (2018), a footbridge is made up of a number of printed mortar voussoirs connected by prestressing cables.

6.2.2. *Mortar reinforcement*

A natural idea, of course, is to integrate the reinforcement into the cementitious material right from the initial mixing stage and extrude fiber-reinforced concretes as described above. To obtain a pumpable mortar, the fibers must be flexible and therefore have a limited diameter. The most commonly used fibers are synthetic (carbon, PVA and PE) or mineral (glass or basalt). In order to maintain a mode of rupture by interface decohesion and/or to avoid blocking the pumping and extrusion system, the length of the fibers is also limited ($\approx$ 10mm). Finally, to guarantee a sufficiently low yield stress, the volume proportion is limited, generally less than or equal to 2% (Ogura et al. 2018).

An interesting result concerns the orientation of fibers during extrusion (Ma et al. 2019; Panda et al. 2017; Hambach and Volkmer 2017; Hambach et al. 2016; Li et al. 2020), which significantly increases reinforcement in the direction of extrusion. The study by Ma et al. (2019), on the other hand, suggests that interlayer interface strength is reduced when reinforcing fibers are added. As with cast-reinforced concretes, depending on the dimensions and fiber content, the mechanical strength of the reinforced material may be improved but the ductility reduced (Hambach et al. 2016), or it may be highly ductile with reduced mechanical strength (Soltan and Li 2018; Ogura et al. 2018; Zhu et al. 2019). The opportunity to obtain this type of ductile, positive-hardening behavior with a mortar, is very interesting for construction and it enriches the family of ECC mentioned above for cast solutions (Li et al. 2020).

The mechanical strengths obtained in the extrusion direction can be higher than those of cast composites (Soltan and Li 2018; Ogura et al. 2018), thanks to a phenomenon of increased multi-cracking in the deposition direction. In Zhu et al. (2019), a PE fiber volume ratio of 2% enabled a very high strain-to-failure in the deposition direction of the order of 10% for a flexural strength of 20 MPa. A similar formulation was recently used on a large scale to produce a 1.5-m column (Li et al.

2020). Earlier work (Hambach et al. 2016) had achieved flexural strengths of 100 MPa with 3% carbon fiber, albeit with a very high yield stress making extrusion difficult.

It is highly likely that, through optimal choices of fiber type and size, mortar quality and process suitability, the effectiveness of this type of reinforcement for printed mortars will improve significantly in the years to come.

6.2.3. *In-line reinforcements*

The idea here is to concomitantly print the mortar and place the reinforcement. These approaches are more sophisticated and generally developed for nonstandard concrete structures and as part of an integrated digital process based on advanced robotic manufacturing strategies.

A number of research projects have focused on the direct adaptation of the principle of reinforced concrete using steel bars, or *rebars* (see, for example, Classen et al. (2020); Perrot et al. (2020); Hass et al. (2022)). This is an interesting approach as it offers the prospect of rapid acceptance and use by the industry. Most solutions involve the robotic insertion of small bars, rather like nails, into the printed stack, either perpendicular to the stack direction or at an angle. This has the advantage of also reinforcing the interfaces crossed, as these can represent a weakness (*cold joints*) for the structure if there is poor adhesion between the layers due to too rapid drying of the layer that will receive the new deposit.

Classen et al. (2020) proposed such a process called additive manufacturing of reinforced concrete (AMoRC), a hybrid process consisting of an intermittent stud welding process and a continuous concrete extrusion process. Segmented steel reinforcing bars are assembled to form a three-dimensional reinforcing mesh and simultaneously embedded in the extruded concrete (Figure 6.6).

In Perrot et al. (2020), experiments with this type of nail reinforcement on different mortars, here already printed, demonstrated effective reinforcement if the orientation of the nails is correctly chosen and if the surface of the nails is sufficiently rough to ensure a good interface with the mortar (Figure 6.7).

In Geneidy et al. (2020), U-shaped reinforcing elements, or "staples", staple the layers together vertically during printing. The staples penetrate several layers, but also interlock to form a reinforcing network. Finally, in Gebhard et al. (2020), a device incorporating a reel enables needles to be deposited horizontally between the layers in order to obtain reinforcement in the axis of the strands, thus approximating to conventional reinforced concrete.

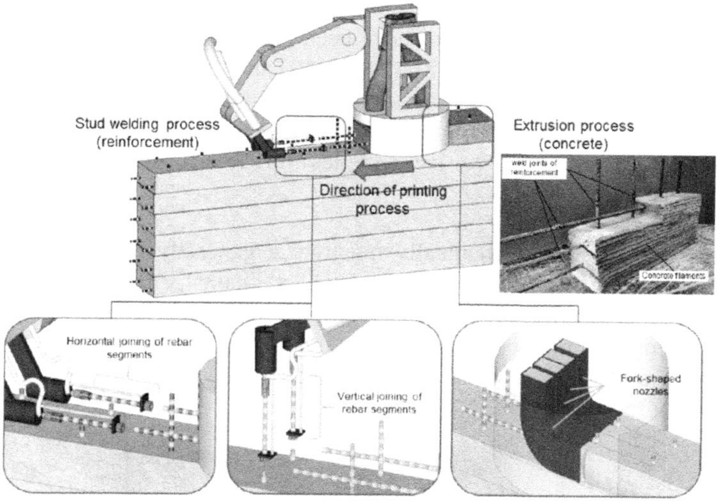

Figure 6.6. *3D printing process (AMoRC) for reinforced concrete, concrete extrusion and reinforcement welding (Classen et al. 2020). For a color version of this figure, see www.iste.co.uk/perrot/3dconcrete.zip*

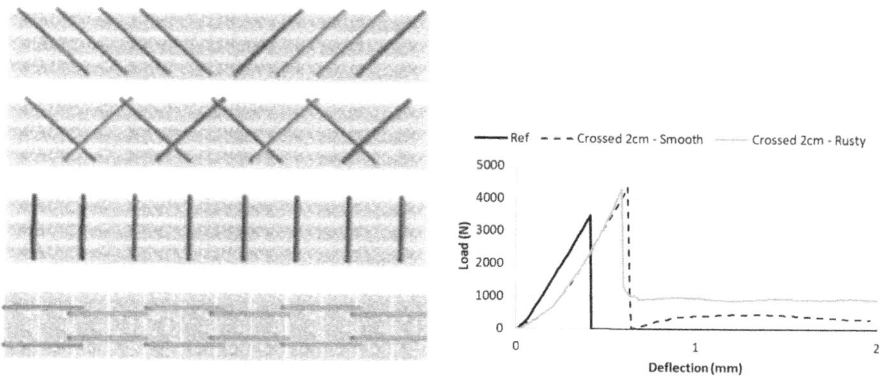

Figure 6.7. *Reinforcement with nails between layers and bending curves showing improved ductility (Perrot et al. 2020)*

Smart dynamic casting and *mesh mould* developed at ETH Zurich (Lloret et al. 2015) offer another robotic concept that enables the continuous production of a three-dimensional welded reinforcing mesh, used here as formwork, and the concrete filler. The technology is based on the ability of this porous cage to retain the mortar, which has sufficient yield stress and viscosity properties while mitigating hydrostatic pressure. Such sacrificial filter forms have been proposed for conventional concrete in France (*Dipy* 1992, *3DR* 1999 (SF3DR 1999)) and then in China (Ren and Wang 2005). They allow for complex shapes and hydrostatic pressure reduction of over 80%. The Empa's NEST building in Zurich is a showcase for experiments using these new 3D-printed concrete technologies for free-form columns and roofs produced without formwork.

Another option is to mechanically integrate a generally continuous, unwound reinforcement at the nozzle, such as micro-cables, chains or metal grids (Bos et al. 2019; Lim et al. 2018; Bos et al. 2017; Marchment and Sanjayan 2020) (Figure 6.8). Particular attention must be paid to the quality of the interface between the reinforcement and the cementitious material, which determines the composite performance of the assembly.

Figure 6.8. *Co-extrusion of reinforced strand, reinforced by continuous reinforcement: technical devices for a metal chain (Bos et al. 2019). For a color version of this figure, see www.iste.co.uk/perrot/3dconcrete.zip*

Mineral fiber elements are of course widely investigated, for example, recently in Mechtcherine et al. (2020a) or by the authors of this chapter in Ducoulombier et al. (2020). The former uses a thin, narrow strip of *mineral-impregnated carbon fiber* (MCF) unwound downstream of the nozzle onto an already printed layer, which is covered by the current deposit. In Mechtcherine et al. (2020b), to ensure proper impregnation of the carbon fiber roving, a pre-impregnation line using very fine mortar is set up upstream to prepare the web to be deposited. The wires are then placed mechanically between the layers upstream of the printed strand.

A variant of the MCF process called *ProfiCarb* (Neef and Mechtcherine 2022) reconditions prepreg yarns into coils that are driven directly by the flow of a single-component mortar (Figure 6.9). Several reinforcements, in this case six yarns of twisted carbon fiber, can then be dispersed in the cross-section of the coil (Figure 6.9).

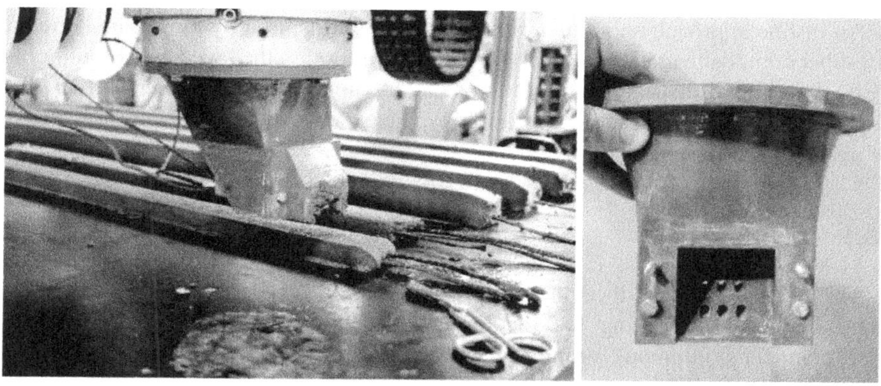

Figure 6.9. *ProfiCarb process and view of the six yarns nozzle (Ivaniuk et al. 2022; Neef and Mechtcherine 2022). For a color version of this figure, see www.iste.co.uk/perrot/3dconcrete.zip*

This latter process is very close to the flow-based-pultrusion (FBP) concept proposed by the authors of this chapter (Ducoulombier et al. 2020) and which will be detailed in section 6.3. This type of approach enables more homogeneous, "in-line" reinforcement and avoids technologies and motors for guiding and driving the reinforcement, which complicate the process and reduce formal possibilities, such as pronounced curves for example.

Last but not least, we should mention the *Injection 3D Concrete Printing* (I3DCP) process (Hack et al. 2020), which enables the injection of one or more continuous reinforcements. This does not involve 3D concrete printing by extrusion, but rather printing in a yield stress fluid bath serving as a support.

All of these continuous reinforcement solutions are, of course, similar in spirit to traditional reinforced concrete techniques. They reinforce preferentially in a preferred direction (Asprone et al. 2018b), induce anisotropy and are therefore very effective for the bending behavior of the structure, even for moderate reinforcement volume fractions. To go further, they need to fit as closely as possible into the framework of formal freedom offered by robotics.

The following paragraph focuses on the FBP concept (Ducoulombier et al. 2020; Caron et al. 2021), which we believe meets the need for efficiency, simplicity and adaptability to new building system proposals.

6.3. Details of a special in-line reinforcement, the flow-based-pultrusion concept

An in-line approach and a material track strongly inspired by the technologies of long-fiber composite materials, since it involves reinforcing the layers with continuous fibers aligned and distributed in a homogeneous, nonlocalized manner, is therefore proposed. We briefly describe the technology, the experiments and the performance of the reinforced material.

6.3.1. *Technology*

The technology described in the patent by Caron and Ducoulombier (2020) is called FBP. By controlling the rheological behavior of the cementitious matrix, small-diameter continuous strands (glass, basalt, etc.) are routed and impregnated without any motorization (Figure 6.10). The resulting material, known as anisotropic concrete, is homogeneously reinforced in a single direction, opening up new possibilities. Indeed, it cannot only improve the strength and ductility of the hardened material, but also contribute to better handling of fresh beads during deposition, enabling more complex sloping or cantilevered courses, as we shall see later (Figure 6.12, right).

6.3.2. *Prototypes and devices*

A first prototype was developed at the Navier laboratory and was adapted to a bicomponent head from XtreeE mounted on an ABB 6200 six-axis robot (Figure 6.11).

As seen in Chapter 2, in concrete 3D printing by extrusion, there are two main technologies, mono-component or bicomponent. The first deposits a layer directly pumped into a batch of prepared mortar of the right consistency for stacking. A powerful pump is required, as the mixture is relatively thin (a few thousand kPa), a yield stress that evolves quite slowly but irremediably to a consistency too high for printing. Two-component technology prints a much more fluid material (a few hundred kPa), strongly accelerated by an additive just before printing. The initial mixture is therefore easier to pump, more stable and allows greater freedom, since the operator can modify it continuously according to what is required or needs to be corrected. Our FBP process is closely linked to this technology, which makes it easy to introduce fibers into a fluid material that hardens rapidly. Impregnation is highly effective and drive remains assured as the mortar quickly builds-up itself around the fiber. First prints have been made with glass, basalt and carbon fibers, up to 6% by volume. Details of the process technology can be found in Demont et al. (2021). To take things further, a second industrial prototype is currently under development in collaboration with XtreeE.

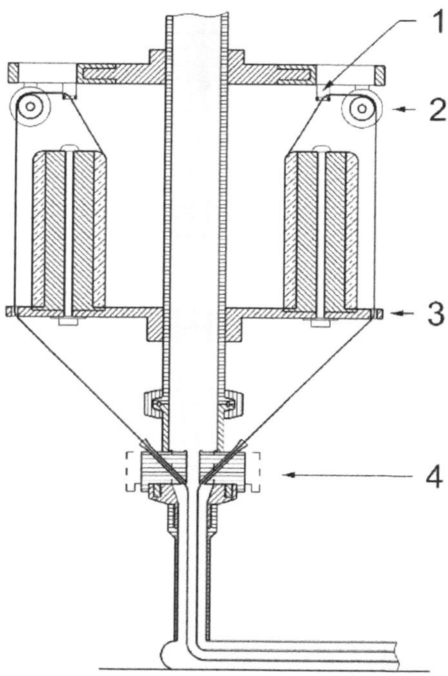

Figure 6.10. *Fiberized printing device: (1) coils, (2) pulleys, (3) guides and (4) drive head (Caron and Ducoulombier 2020)*

Figure 6.12, the XtreeE prototype, shows the cross-section of a hardened bead reinforced with 1% carbon fiber in the center. On the right, it shows a corolla that would not be printable without fiber, as the rings are rapidly subjected to tensile stresses that mortar alone cannot withstand.

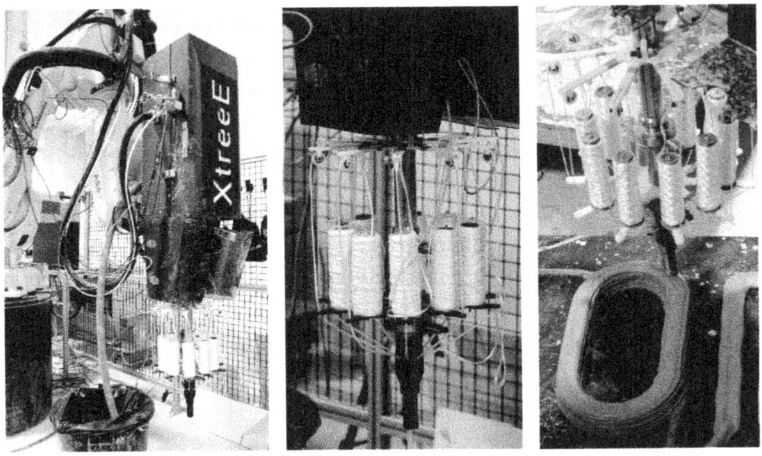

(a) (b) (c)

Figure 6.11. *Prototype 1, Navier laboratory, with fiberglass printing. For a color version of this figure, see www.iste.co.uk/perrot/3dconcrete.zip*

Figure 6.12. *Prototype 2, XtreeE: printing with carbon fibers, evenly distributed as shown by the cross-section of strands in the center. Printing of a fibrous structure with tensile rings (right). For a color version of this figure, see www.iste.co.uk/perrot/3dconcrete.zip*

6.3.3. *Characteristics of the hardened and reinforced material*

Tensile and flexural tests have been carried out on cords reinforced with various levels of glass fiber and basalt. Other tests are in progress with other fibers such as carbon or PVA. The specimens consist of a tensile layer and a flexural multi-layer. Figure 6.13 shows the type of tensile behavior obtained with 3% fibers (in this case glass), and the high ductility achieved thanks to the diffuse microcracking made possible by the presence of the fibers. This type of behavior is also found in bending, and details can be found in Caron et al. (2021). Meter-scale tests on beams with reinforced intrados are currently underway, and initial results are very promising.

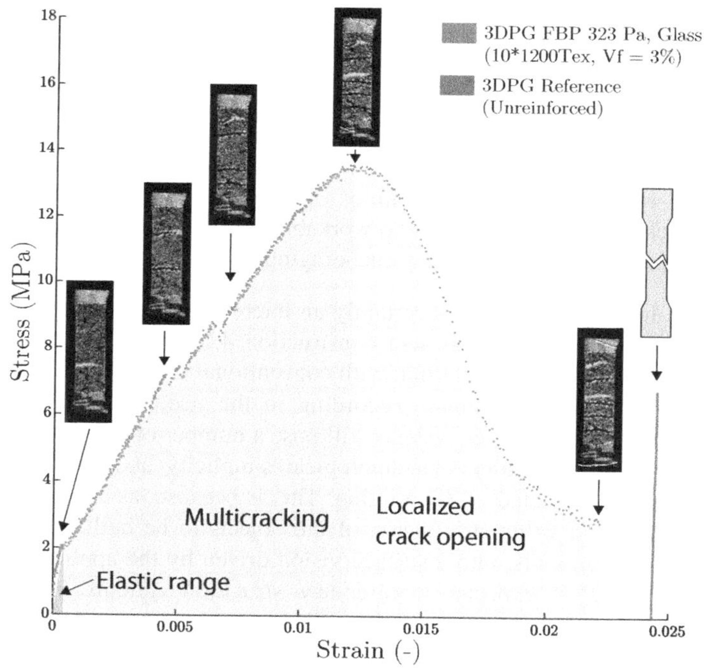

Figure 6.13. *Tensile test of a 3% glass fiber-reinforced specimen and image correlation revealing the progressive appearance of cracks. On the left, in gray, an unreinforced specimen. For a color version of this figure, see www.iste.co.uk/perrot/3dconcrete.zip*

6.4. Conclusion and outlook

Due to the low mechanical properties obtained in tension and the regulation framework, reinforcing 3D mortar prints is an imperative and a highly strategic research and industrial issue for the commercial development of the technology.

While there are no off-the-shelf solutions today, a number of initiatives are already providing answers to the various situations encountered. Often inspired by the long and rich history of reinforcement technologies for cast cements and concretes, they also try to draw on the new tools available today, such as robotics, parametric design and materials science.

In the first part, we briefly recalled the important elements in the history of "reinforced concrete", the material mechanisms involved and the challenges in terms of strength and ductility control. These are concentrated in the term "engineered cementitious composite", in other words the formulation of tailor-made cementitious materials.

In the second part, we take stock of existing initiatives, and classify them according to whether they reenforce the printed structure, the printing material, or in-line. The former have the advantage of being able to be rapidly implemented in real-life situations, via demonstrators, and thus promote the technology. The second type is very promising, as this type of fiber-reinforced mortar is already widely accepted and regulated in the community. Technical progress is still needed to improve the achievable fiber content and workability of mortars, but initial results in terms of strength and ductility are very encouraging.

Finally, in-line reinforcement is certainly an interesting avenue, perhaps more in tune with the spirit of technology and construction 4.0. This should enable more massive reinforcements, in competition with conventional steel systems for concrete, more flexible in use and adaptable according to the need and the object to be reinforced. However, these latter systems still pose a number of problems: anchoring and flexibility of reinforcements, technological simplicity and suitability for the formal performance expected of 3D printing. This is because, as already mentioned, the latter transforms a volumetric vision of the objects to be built, induced by the formwork and casting tools, into a surface vision driven by the application process. 3D printing is therefore expected to offer new structural systems. We cite here a particular example of a concrete lattice, illustrated in Figure 6.14, a concept presented in Duballet et al. (2019) and realized via robotized manufacturing by free deposition (Duballet et al. 2020). In this type of structure, the elements only work in tension or compression. The anisotropy provided by an in-line robotized fiber extrusion process would therefore be particularly suitable and totally disruptive in terms of concrete construction systems.

In the final section, we focus on an initiative by the authors of this chapter, that is, FBP. This type of in-line reinforcement could meet several key challenges: good anchorage of the fibers in the cementitious matrix, technological simplicity and a source of inspiration for new construction systems.

In conclusion, as mentioned in the introduction, the competition between two reinforcement systems proposed at the end of the 19th century – the *Hennebique* system (reinforced posts and beams) and the *Cottencin* system (thin ferrocement slabs) – could well be replayed in favor of the latter, thanks to new printing processes and in-line reinforcement as proposed by FBP. This would make it possible to reinvent freer, lighter structures such as those proposed by Candela or Pier Luigi Nervi (Figure 6.1). In his own words, Nervi wanted to "get away from plank architecture" by eliminating the need for wooden formwork, and serve a vision that combined material, form and structure to create more virtuous buildings.

Figure 6.14. *Curved concrete truss presented at the "FORM and FORCE" international conference in Barcelona in October 2019. (Photo credit: Stephano Borghi). For a color version of this figure, see www.iste.co.uk/perrot/3dconcrete.zip*

6.5. References

Allen, M.D. and Raabe, O.G. (1985). Slip correction measurements of spherical solid aerosol particles in an improved millikan apparatus. *Aerosol Science and Technology*, 4(3), 269–286.

Asprone, D., Auricchio, F., Menna, C., Mercuri, V. (2018a). 3D printing of reinforced concrete elements: Technology and design approach. *Construction and Building Materials*, 165, 218–231 [Online]. Available at: https://linkinghub.elsevier.com/retrieve/pii/S0950061818300187.

Asprone, D., Menna, C., Bos, F.P., Salet, T.A., Mata-Falcón, J., Kaufmann, W. (2018b). Rethinking reinforcement for digital fabrication with concrete. *Cement and Concrete Research*, 112, 111–121 [Online]. Available at: https://linkinghub.elsevier.com/retrieve/pii/S0008884618300309.

Betterman, L.R. and Ouyang, C. (1995). Fiber-matrix interaction in microfiber-reinforced mortar. *Advanced Cement Based Materials*, 2, 53–61.

Bos, F.P., Ahmed, Z., Jutinov, E., Salet, T.A.M. (2017). Experimental exploration of metal cable as reinforcement in 3D printed concrete. *Materials*, 10(11), 1314 [Online]. Available at: http://www.mdpi.com/1996-1944/10/11/1314.

Bos, F.P., Bosco, E., Salet, T.A.M. (2019). Ductility of 3D printed concrete reinforced with short straight steel fibers. *Virtual and Physical Prototyping*, 14(2), 160–174. doi: 10.1080/17452759.2018.1548069.

Brandt, A.M. (2008). Fibre reinforced cement-based (FRC) composites after over 40 years of development in building and civil engineering. *Composite Structures*, 86(1–3), 3–9 [Online]. Available at: https://linkinghub.elsevier.com/retrieve/pii/S0263822308000597.

Brückner, A., Ortlepp, R., Curbach, M. (2006). Textile reinforced concrete for strengthening in bending and shear. *Materials and Structures*, 39(8), 741–748 [Online]. Available at: http://link.springer.com/10.1617/s11527-005-9027-2.

Carneau, P., Mesnil, R., Roussel, N., Baverel, O. (2019). An exploration of 3D printing design space inspired by masonry. In *IASS Symposium 2019, "Form and Force"*, October, Barcelona.

Carneau, P., Mesnil, R., Roussel, N., Baverel, O. (2020). Additive manufacturing of cantilever – From masonry to concrete 3D printing. *Automation in Construction*, 116, 103184 [Online]. Available at: https://linkinghub.elsevier.com/retrieve/pii/S0926580519308568.

Caron, J.-F, and Ducoulombier, N. (2020). Method and device for manufacturing anisotropic concrete. Patent, CN114174029B.

Caron, J.-F., Demont, L., Ducoulombier, N., Mesnil, R. (2021). 3D printing of mortar with continuous fibres: Principle, properties and potential for application. *Automation in Construction*, 129, 103806 [Online]. Available at: https://www.sciencedirect.com/science/article/pii/S0926580521002570.

Castelier, E., Gélébart, L., Lacour, C., Lantuéjoul, C. (2010). Three consistent approaches of the multiple cracking process in 1D composites. *Composites Science and Technology*, 70(15), 2146–2153 [Online]. Available at: https://linkinghub.elsevier.com/retrieve/pii/S0266353810003222.

Chanvillard (1992). Analyse expérimentale et modélisation micromécanique du comportement des fibres d'acier tréfilées, ancrées dans une matrice cimentaire. PhD Thesis, Université de Sherbrooke, Sherbrooke.

Chateau, C., Gélébart, L., Bornert, M., Crépin, J. (2015). Micromechanical modeling of the elastic behavior of unidirectional CVI SiC/SiC composites. *International Journal of Solids and Structures*, 58, 322–334 [Online]. Available at: https://linkinghub.elsevier.com/retrieve/pii/S0020768314004478.

Chudoba, R., Vořechovský, M., Konrad, M. (2006). Stochastic modeling of multi-filament yarns. I. Random properties within the cross-section and size effect. *International Journal of Solids and Structures*, 43(3–4), 413–434 [Online]. Available at: https://linkinghub.elsevier.com/retrieve/pii/S0020768305004087.

Classen, M., Ungermann, J., Sharma, R. (2020). Additive manufacturing of reinforced concrete – Development of a 3D printing technology for cementitious composites with metallic reinforcement. *Applied Sciences*, 10(11) [Online]. Available at: https://www.mdpi.com/2076-3417/10/11/3791.

Curtin, W., Ahn, B., Takeda, N. (1998). Modeling brittle and tough stress – Strain behavior in unidirectional ceramic matrix composites. *Acta Materialia*, 46(10), 3409–3420 [Online]. Available at: https://linkinghub.elsevier.com/retrieve/pii/S135964549800041X.

Demont, L., Ducoulombier, N., Mesnil, R., Caron, J.-F. (2021). Flow-based pultrusion of continuous fibers for cement-based composite material and additive manufacturing: Rheological and technological requirements. *Composite Structures*, 262, 113564 [Online]. Available at: https://www.sciencedirect.com/science/article/pii/S0263822321000258.

Duballet, R. (2018). Building systems in robotic extrusion of cementitious materials. PhD Thesis, Université Paris-Est Créteil, Paris.

Duballet, R., Baverel, O., Dirrenberger, J. (2019). Space truss masonry walls with robotic mortar extrusion. *Structures*, 18, 41–47 [Online]. Available at: https://linkinghub.elsevier. com/retrieve/pii/S2352012418301309.

Duballet, R., Mesnil, R., Ducoulombier, N., Carneau, P., Demont, L., Motamedi, M., Baverel, O., Caron, J.-F., Dirrenberger, J. (2020). Free deposition printing for space truss structures. In *Second RILEM International Conference on Concrete and Digital Fabrication*, Bos, F., Lucas, S., Wolfs, R., Salet, T. (eds). Springer, Cham.

Ducoulombier, N., Demont, L., Chateau, C., Bornert, M., Caron, J.-F. (2020). Additive manufacturing of anisotropic concrete: A flow-based pultrusion of continuous fibers in a cementitious matrix. *Procedia Manufacturing*, 47, 1070–1077 [Online]. Available at: https://www.sciencedirect.com/science/article/pii/S2351978920311732.

Gebhard, L., Mata-Falcón, J., Anton, A., Burger, J., Lloret-Fritschi, E., Reiter, L., Dillenburger, B., Gramazio, F., Kohler, M., Flatt, R., Kaufmann, W. (2020). Aligned interlayer fibre reinforcement and post-tensioning as a reinforcement strategy for digital fabrication. In *Second RILEM International Conference on Concrete and Digital Fabrication*, Bos, F.P., Lucas, S.S., Wolfs, R.J., Salet, T.A. (eds). Springer, Cham.

Geneidy, O., Kumarji, S., Dubor, A., Sollazzo, A. (2020). Simultaneous reinforcement of concrete while 3D printing. In *Second RILEM International Conference on Concrete and Digital Fabrication*, Bos, F.P., Lucas, S.S., Wolfs, R.J., Salet, T.A. (eds). Springer, Cham.

Hack, N., Dressler, I., Brohmann, L., Gantner, S., Lowke, D., Kloft, H. (2020). Injection 3D concrete printing (I3DCP): Basic principles and case studies. *Materials*, 13(5), 1093.

Hambach, M. and Volkmer, D. (2017). Properties of 3D-printed fiber-reinforced Portland cement paste. *Cement and Concrete Composites*, 79, 62–70 [Online]. Available at: https://linkinghub.elsevier.com/retrieve/pii/S0958946516301093.

Hambach, M., Müller, H., Neumann, T., Volkmer, D. (2016). Portland cement paste with aligned carbon fibers exhibiting exceptionally high flexural strength (> 100 MPa). *Cement and Concrete Research*, 89, 80–86 [Online]. Available at: https://linkinghub.elsevier.com/retrieve/pii/S0008884616302630.

Hamidi, F. and Aslani, F. (2019). Additive manufacturing of cementitious composites: Materials, methods, potentials, and challenges. *Construction and Building Materials*, 218, 582–609 [Online]. Available at: https://linkinghub.elsevier.com/retrieve/pii/S0950061819313194.

Hass, L., Bos, F.P., Salet, T.A.M. (2022). Characterizing the bond properties of automatically placed helical reinforcement in 3D printed concrete. *Construction and Building Materials*, 355, 129228.

Häußler-Combe, U. and Hartig, J. (2007). Bond and failure mechanisms of textile reinforced concrete (TRC) under uniaxial tensile loading. *Cement and Concrete Composites*, 29(4), 279–289 [Online]. Available at: http://linkinghub.elsevier.com/retrieve/pii/S0958946507000091.

Hegger, J. and Voss, S. (2008). Investigations on the bearing behaviour and application potential of textile reinforced concrete. *Engineering Structures*, 30(7), 2050–2056 [Online]. Available at: http://linkinghub.elsevier.com/retrieve/pii/S0141029608000126.

Hegger, J., Will, N., Bruckermann, O., Voss, S. (2006). Load–bearing behaviour and simulation of textile reinforced concrete. *Materials and Structures*, 39(8), 765–776 [Online]. Available at: http://link.springer.com/10.1617/s11527-005-9039-y.

Hegger, J., Curbach, M., Stark, A., Wilhelm, S., Farwig, K. (2018). Innovative design concepts: Application of textile reinforced concrete to shell structures. *Structural Concrete*, 19(3), 637–646. doi: 10.1002/suco.201700157.

Ivaniuk, E., Müller, S., Neef, T., Mechtcherine, V. (2022). Strategies for integrating reinforcement into 3D concrete printing at the TU Dresden. *Open Conference Proceedings*, 1, 23–34.

Jun, P. and Mechtcherine, V. (2010). Behaviour of strain-hardening cement-based composites (SHCC) under monotonic and cyclic tensile loading: Part 2 – Modelling. *Cement and Concrete Composites*, 32(10), 810–818 [Online]. Available at: http://linkinghub.elsevier.com/retrieve/pii/S0958946510001162.

Keita, E., Bessaies-Bey, H., Zuo, W., Belin, P., Roussel, N. (2019). Weak bond strength between successive layers in extrusion-based additive manufacturing: Measurement and physical origin. *Cement and Concrete Research*, 123, 105787 [Online]. Available at: https://linkinghub.elsevier.com/retrieve/pii/S0008884618313760.

Khoshnevis, B. (2004). Automated construction by contour crafting–related robotics and information technologies. *Automation in Construction*, 13(1), 5–19 [Online]. Available at: https://linkinghub.elsevier.com/retrieve/pii/S0926580503000736.

Kouris, L.A.S. and Triantafillou, T.C. (2018). State-of-the-art on strengthening of masonry structures with textile reinforced mortar (TRM). *Construction and Building Materials*, 188, 1221–1233 [Online]. Available at: https://linkinghub.elsevier.com/retrieve/pii/S0950061818319792.

Li, V.C. (2003). On engineered cementitious composites (ECC). *Journal of Advanced Concrete Technology*, 1(3), 215–230.

Li, V.C. and Leung, C.K.Y. (1992). Steady–state and multiple cracking of short random fiber composites. *Journal of Engineering Mechanics*, 118(11), 2246–2264.

Li, V.C., Bos, F.P., Yu, K., McGee, W., Ng, T.Y., Figueiredo, S.C., Nefs, K., Mechtcherine, V., Nerella, V.N., Pan, J. et al. (2020). On the emergence of 3D printable engineered, strain hardening cementitious composites (ECC/SHCC). *Cement and Concrete Research*, 132, 106038 [Online]. Available at: https://linkinghub.elsevier.com/retrieve/pii/S0008884620300557.

Lim, S., Buswell, R., Le, T., Austin, S., Gibb, A., Thorpe, T. (2012). Developments in construction-scale additive manufacturing processes. *Automation in Construction*, 21, 262–268 [Online]. Available at: https://linkinghub.elsevier.com/retrieve/pii/S0926580511001221.

Lim, J.H., Panda, B., Pham, Q.-C. (2018). Improving flexural characteristics of 3D printed geopolymer composites with in-process steel cable reinforcement. *Construction and Building Materials*, 178, 32–41 [Online]. Available at: https://linkinghub.elsevier.com/retrieve/pii/S0950061818310675.

Lloret, E., Shahab, A., Linus, M., Flatt, R., Gramazio, F., Kohler, M., Langenberg, S. (2015). Complex concrete structures: Merging existing casting techniques with digital fabrication. *Computer-Aided Design*, 60, 40–49.

Ma, G., Li, Z., Wang, L., Wang, F., Sanjayan, J. (2019). Mechanical anisotropy of aligned fiber reinforced composite for extrusion-based 3D printing. *Construction and Building Materials*, 202, 770–783 [Online]. Available at: https://linkinghub.elsevier.com/retrieve/pii/S095006181930008X.

Maitenaz, S., Charrier, M., Mesnil, R., Onfroy, P., Metge, N., Feraille, A., Caron, J.-F. (2021). Fabrication of a truss-like beam casted with 3D printed clay moulds. In *Proceedings of the 38th International Symposium on Automation and Robotics in Construction (ISARC)*, Feng, C., Linner, T., Brilakis, I.. Castro, D., Chen, P.-H., Cho, Y., Du, J., Ergan, S., Garcia de Soto, B., Gašparik, J. et al. (eds). IAARC, Munich.

Maitenaz, S., Mesnil, R., Onfroy, P., Metge, N., Feraille, A., Caron, J.-F. (2024). Materialising structural optimisation of reinforced concrete beams through digital fabrication. *Structures*, 59, 105644. doi: 10.1016/j.istruc.2023.105644.

Marchment, T. and Sanjayan, J. (2020). Mesh reinforcing method for 3D concrete printing. *Automation in Construction*, 109, 102992 [Online]. Available at: https://linkinghub.elsevier.com/retrieve/pii/S0926580519306132.

Martinie (2010). Comportement rhéologique et mise en œuvre des matériaux cimentaires fibrés. PhD Thesis, Université Paris-Est Créteil, Paris.

Mechtcherine, V. (2013). Novel cement-based composites for the strengthening and repair of concrete structures. *Construction and Building Materials*, 41, 365–373 [Online]. Available at: https://linkinghub.elsevier.com/retrieve/pii/S0950061812009555.

Mechtcherine, V., Michel, A., Liebscher, M., Schmeier, T. (2020a). Extrusion-based additive manufacturing with carbon reinforced concrete: Concept and feasibility study. *Materials*, 13(11) [Online]. Available at: https://www.mdpi.com/1996-1944/13/11/2568.

Mechtcherine, V., Michel, A., Liebscher, M., Schneider, K., Großmann, C. (2020b). Mineral-impregnated carbon fiber composites as novel reinforcement for concrete construction: Material and automation perspectives. *Automation in Construction*, 110, 103002.

Naaman, A.E. (2000). *Ferrocement and Laminated Cementitious Composites*, 1st edition. Techno Press, Ann Arbor.

Neef, T. and Mechtcherine, V. (2022). Simultaneous integration of continuous mineral-bonded carbon reinforcement into additive manufacturing with concrete. *Open Conference Proceedings*, 1, 73–81.

Ogura, H., Nerella, V.N., Mechtcherine, V. (2018). Developing and testing of strain-hardening cement-based composites (SHCC) in the context of 3D-printing. *Materials*, 11(8), 1375.

Panda, B., Chandra Paul, S., Jen Tan, M. (2017). Anisotropic mechanical performance of 3D printed fiber reinforced sustainable construction material. *Materials Letters*, 209, 146–149 [Online]. Available at: http://linkinghub.elsevier.com/retrieve/pii/S0167577X17311679.

Pegna, J. (1997). Exploratory investigation of solid freeform construction. *Automation in Construction*, 5(5), 427–437 [Online]. Available at: https://linkinghub.elsevier.com/retrieve/pii/S0926580596001665.

Peled, A. and Mobasher, B. (2006). Properties of fabric – Cement composites made by pultrusion. *Materials and Structures*, 39(8), 787–797 [Online]. Available at: http://link.springer.com/10.1617/s11527-006-9171-3.

Peled, A., Bentur, A., Yankelevsky, D. (1994). Woven fabric reinforcement of cement matrix. *Advanced Cement Based Materials*, 1(5), 216–223 [Online]. Available at: http://linkinghub.elsevier.com/retrieve/pii/1065735594900272.

Perrot, A., Jacquet, Y., Rangeard, D., Courteille, E., Sonebi, M. (2020). Nailing of layers: A promising way to reinforce concrete 3D printing structures. *Materials*, 13(7) [Online]. Available at: https://www.mdpi.com/1996-1944/13/7/1518.

Ren, Z. and Wang, L. (2005). Applied research of light steel rib-concrete with dipy construction formwork system. In *Proceedings of the Fourth International Conference on Advances in Steel Structures*. Elsevier, Amsterdam.

Rossi, P. (1998). *Les bétons de fibres métalliques*, 1st edition. Presse des Ponts, Paris.

Rossi, P. (2001). Ultra-high-performance fiber-reinforced concretes. *Concrete International*, 23(12), 46–52.

Rossi, P. (2013). Influence of fibre geometry and matrix maturity on the mechanical performance of ultra high-performance cement-based composites. *Cement and Concrete Composites*, 37, 246–248 [Online]. Available at: https://linkinghub.elsevier.com/retrieve/pii/S0958946512001692.

Salet, T.A.M., Ahmed, Z.Y., Bos, F.P., Laagland, H.L.M. (2018). Design of a 3D printed concrete bridge by testing. *Virtual and Physical Prototyping*, 13(3), 222–236. doi: 10.1080/17452759.2018.1476064.

SF3DR (1999). 3DR – Structural formwork [Online]. Available at: https://www.3drcoffrages.com/en/.

Soltan, D.G. and Li, V.C. (2018). A self-reinforced cementitious composite for building-scale 3D printing. *Cement and Concrete Composites*, 90, 1–13 [Online]. Available at: https://linkinghub.elsevier.com/retrieve/pii/S095894651730536X.

Wille, K., El-Tawil, S., Naaman, A. (2014). Properties of strain hardening ultra high performance fiber reinforced concrete (UHP-FRC) under direct tensile loading. *Cement and Concrete Composites*, 48, 53–66 [Online]. Available at: https://linkinghub.elsevier.com/retrieve/pii/S0958946514000055.

Xia, Z. and Curtin, W. (2000). Tough-to-brittle transitions in ceramic-matrix composites with increasing interfacial shear stress. *Acta Materialia*, 48(20), 4879–4892 [Online]. Available at: https://linkinghub.elsevier.com/retrieve/pii/S1359645400002913.

Zhu, B., Pan, J., Nematollahi, B., Zhou, Z., Zhang, Y., Sanjayan, J. (2019). Development of 3D printable engineered cementitious composites with ultra-high tensile ductility for digital construction. *Materials & Design*, 181, 108088 [Online]. Available at: https://linkinghub.elsevier.com/retrieve/pii/S026412751930526X.

7

Numerical Simulation Tools for 3D Printing

Duc-Phi DO, Zeinab DIAB, Sébastien RÉMOND and Dashnor HOXHA

Lamé, Université d'Orléans, France

7.1. Introduction

Widely developed over the last two decades, additive manufacturing (AM) through the three-dimensional concrete printing (3DCP) of structural elements represents an important step in the automation process of the construction industry. This innovative technique makes it possible to build structures with a high degree of geometric freedom, saving materials and reducing costs and delivery times, compared with traditional construction methods.

In the 3D printing process, the preparation of the geometry of the printed structure, the choice of the material formulation (especially cementitious) and the configuration of the printer with the process parameters are an essential first step before manufacturing. According to this procedure, the digital prototype based on computer-aided design (CAD) of the printed structure is a prerequisite. This virtual geometric model is then transformed into a standard format and cut into layers before being manufactured by the printer. Implementing the programs that control the trajectories and speed of movement of the printer (e.g. the robots) in the digital tools is the primary task in this preparation stage.

3D Concrete Printing,
coordinated by Arnaud PERROT and Yohan JACQUET. © ISTE Ltd 2025.

Due to the novelty of this technique, there is still a lack of knowledge concerning the specific relationships between design, material and process parameters. Indeed, these parameters interact strongly and significantly influence the quality of the manufactured product. The first attempt to find the optimum 3D printing parameters for a given design is based on a trial-and-error approach using a range of print speeds, printed layer dimensions and material composition. However, as soon as the design changes, or new materials become available, this costly procedure has to be restarted, and the print parameters modified as required. Consequently, such an approach limits the development and application of 3D concrete printing on a large scale in practice. If this layer-by-layer manufacturing technique is to be competitive with concrete casting, it must enable a reduction in manufacturing cost and time, without compromising the geometric freedom and structural safety of printed objects. So, alongside the development of geometric modeling and virtual printing simulation tools, it is necessary to integrate 3D printing process modeling into this numerical approach to predict the behavior of deposited layers and structural failure during fresh concrete printing – a phenomenon frequently observed in several trials. In fact, the ability to anticipate the risk of failure of the printed structure thanks to digital modeling in the design phase means that an appropriate layer size and printing speed can be chosen beforehand. During the printing process, these numerical predictions can be updated to enable the right strategy decisions to be made, as well as updating the printing parameters to maximize the success of this innovative construction technique.

The main aim of this chapter is to present an overview of the tools, approaches and recent results of numerical modeling in the field of 3D concrete printing. After summarizing digital tools for geometric modeling and virtual printing, we will detail the state of the art in digital modeling of the 3D printing process. This final section focuses on both the structure and the deposited layer. Finally, after a discussion of recent advances and existing limitations, some potential avenues are presented to guide future research in this field.

7.2. Designing the geometric model of a virtual object

The AM process for concrete structures involves several stages, from the design of the geometric model of a virtual object to the production of the real object and its finishing for commissioning (Abdulhameed et al. 2019). The first step is to design a model using 3D drafting software, respecting the specifications required for the design of the object to be created. Converted into an STL (stereolithography or standard tessellation language) format, this 3D model is then cut into layers and each layer translated into consecutive motion commands using a specific language such as G-Code, which is familiar from CNC (computer numerical control) milling. This code provides the printer with instructions on the path of movement (or tool trajectory) and the corresponding speeds of the printing process. The second step is to configure the machine and prepare the necessary materials before starting production. The final step is to evaluate the printed object in terms of its visual appearance and performance. If quality requirements are not met, adjustments may need to be made to the digital model, machine or materials.

Digital tools offer a variety of functionalities that facilitate the design of innovative structures through AM. A parametric approach, for example, has been adopted at Loughborough University, enabling great adaptability of the parameters involved (robot movement speed, printed layer size) in the concrete 3D printing process (Wolfs 2015). This approach has proved capable of identifying the relationships between these parameters automatically using the simulated annealing optimization technique. For this purpose, a script is fully generated in Grasshopper, a graphical algorithm editor, which is a plugin for Rhinoceros (McNeel 2010), including all printing parameters (material properties, geometry and loads). This script is then sent to the calculation software, which is controlled by the Python programming language. Using the script created by Grasshopper, Python controls the execution of both the structural analysis during 3D printing and the algorithm for optimizing printing time and the quantity of concrete used. Various research projects have made use of this methodology, such as the work of Lim et al. (2016), where Grasshopper was successfully used to generate curved layered print trajectories and transfer them to the robotic arm to print the designed model.

In the context of the construction sector's move toward digitalization, the integration of BIM (building information modeling) with 3D printing activities has become a challenge for many research efforts. BIM is an in-depth approach to building construction management that covers all phases of the construction lifecycle, from planning to construction and eventually maintenance (Eastman 2011). It encompasses essential information such as geometric data, resources, equipment, materials and manufacturing data (Wu et al. 2016). Consequently, implementing a BIM-based additive construction procedure can significantly reduce turnaround times. For example, it can facilitate the immediate application of design changes to 3D printing in construction (Ding et al. 2019). In addition, BIM's ability to store and organize data relating to material delivery, printer control and finishing operations can be used to fully automate the process (Tay et al. 2017; Teizer et al. 2018).

An optimized approach that streamlines the use of BIM in the 3D printing process is proposed in Anane et al. (2021). The suggested approach consists of a tightly coupled approach from BIM software realized using the Rhino.Inside.Revit plugin (Vandezande et al. 2012). The print script is developed in Rhinoceros using Grasshopper, since it is much more widely used in the digital manufacturing environment. Consequently, the print simulation and command lines are instantly generated with Rhino.Inside.Revit, that is, Rhino running in Revit's memory space. The Speckle plugin can be used for collaboration between different stakeholders in a project and will be used to transmit geometric information and exchange print data in real time between the software it supports. The augmented reality representation of 3D printing simulation could be used by Fologram, a real-time connection between mixed reality devices and Rhino and Grasshopper (Fologram for Rhino and Grasshopper n.d.; https://www.food4rhino.com/en/). The result of this process illustrates the simultaneous visualization of the developed 3D print model in four columns from left to right: in Rhinoceros, Revit, Speckle web viewer and in augmented reality, thanks to Fologram. However, this approach still has certain limitations when it comes to mesh transfer, since 3D printing is based on meshes, which are often data intensive.

7.3. Digital modeling of the 3D printing process

Numerical modeling of the 3D concrete printing process is a complex task, involving the prediction of the behavior of components and/or the entire concrete structure during printing. Indeed, to accurately model the entire printing process, the numerical study must be carried out at two critical levels of the process: the material deposition level at the deposited layer scale and the printed structure level. However, modeling the behavior of fresh concrete during the printing process, whose nature varies from a viscous fluid state (as in the extrusion and material deposition steps) to a solid state after deposition of the printed layers, presents a first obstacle. Furthermore, from a numerical point of view, simulating behavior from the material to the structural scale is very costly and always presents a major challenge for engineers and researchers. Consequently, there is no universal model in the literature to represent the behavior of fresh concrete, and numerical modeling of the 3D printing process is very often carried out separately at each scale. For the purposes of this chapter, we will limit our synthesis to the two scales aimed at predicting structural stability (i.e. buildability) during 3D printing and behavior at the deposited layer scale.

7.3.1. *Simulation of buildability at structural scale*

It is generally recognized that structural failure during 3D concrete printing is controlled by two mechanisms: instability due to elastic buckling or plastic collapse of the unsupported printed structure (Roussel 2018; Suiker 2018; Wolfs et al. 2018; Kruger et al. 2019; Roussel et al. 2020; Suiker et al. 2020). The progressive increase in load due to the self-weight of successive layers in the absence of formwork and the insufficient development of the strength and stiffness properties of extruded concrete throughout the process are the sources of these failure mechanisms (Reiter et al. 2018). The study of buildability aims to predict the possibility of printing the structure before it fails, and initial attempts have been made on the basis of analytical solutions. These analytical solutions were based on comparing the gradual increase in load due to the addition of printed layers, to a failure stress determined from the shear yield stress of the fresh material considered as a yield stress fluid (Perrot et al. 2016; Wangler et al. 2016). In addition to this plastic failure criterion, Roussel (2018) presented an analytical method for determining the critical height at which elastic buckling is expected to occur for a wall subjected solely to its own weight. The advantage of these analytical solutions lies in their ability to easily determine suitable printing parameters (such as robot travel speed, dimensions of the printed structure), thus providing an initial estimate of structural failure prevention requirements during 3D printing. However, the simplicity of this analytical approach means that the gradient of material properties and variations in process parameters over the height of the object cannot be taken into account. To take specific account of these different parameters, it is necessary to use methods based on numerical modeling. Table 7.1 summarizes the various contributions to numerical modeling of the 3D printing process.

The application of numerical modeling to study structural behavior during the 3D printing process was first shown by Wolfs et al. (Reiter et al. 2018). A 3D finite element model was generated for computation by Abaqus software and driven by a custom parametric Python script. This model simulates the successive addition of layers and thus the added gravity load through a nonlinear geometric analysis until completion or failure of the structure. At the interfaces, the surfaces of the layers have been fused together by a geometric bonding constraint, so that there is no relative movement between them. The speed of activation depends on the parameters of the printing process and the length of the object contour. The mechanical properties of the material develop for each layer throughout the analysis as a function of their age in the printing process. In this study, the Mohr–Coulomb (MC)-type plastic failure criterion was adopted and a linear time-dependent evolution of mechanical properties (such as Young's modulus and cohesion) of fresh concrete was suggested. These required material properties were obtained from an experimental set-up comprising a direct shear test and a uniaxial compression test. The numerical predictions of Wolfs et al. (2018) were verified by comparison with experimental results. This model enables structural behavior during 3D printing to be studied systematically, taking into account two failure modes, namely, elastic buckling and plastic collapse. However, from a quantitative point of view, the model overestimated the prediction of the maximum number of printed layers before failure by 58.6%, compared with the experimental results. This difference was explained by the overestimation of material properties determined from testing due to sample compaction, or by neglecting the influence of geometric imperfections in the printed structure. Furthermore, this analysis only gives indications of the object's structural behavior such as the maximum number of printable layers, but no suggestions on how to improve the geometry or determine the best printing parameters that can be provided.

In another study by Wolfs et al. (2019b), geometric imperfections are considered for numerical predictions of structural failures. Indeed, the first buckling mode profile obtained by bifurcation analysis is taken to model the initial imperfection whose amplitude is equal to 0.1% of the deposited layer width. Through analysis of the elasto-plastic behavior of the printed structure, the failure height is defined as that at which the out-of-plane displacement of the structure exceeds half the width of the printed layer. Different walls of the same 60 mm width but varying lengths (L = 1 m, 5 m or 10.4 m) were printed to failure for experimental verification of the numerical model, whose input parameters for the mechanical properties of the concrete were determined by the results of triaxial compression tests. The comparison shows an underestimation of the numerical model of around 15% of the critical height before failure for a 5 m long wall, and the discrepancy can reach 43% for a longer wall (L = 10.4 m). The discrepancy between numerical predictions and printing experiments was explained by the method used to characterize the properties of the printed material, which ignored the presence of confining pressure in the lower regions of large structures, as well as the effect of temperature on material properties. These studies encourage the advancement of numerical modeling of 3D printing in order to increase the accuracy of predictions and take into account thermal and humidity factors due to their significant impact on early-age strength and stiffness properties, and thus on print failures (Wolfs et al. 2019b). These developments could enable numerical assessment of the strength of adhesion between successive layers, which has been shown to be highly dependent on process parameters and environmental conditions (Le et al. 2012; Keita et al. 2019; Nerella et al. 2019; Wolfs and Suiker 2019; Wolfs et al. 2019a). In addition, it could serve as a starting point for extending the simulation to the hardened state.

	Method	Software	Breaking modes	Interface between layers	Imperfections	Time effect	Geometric model
(Wolfs et al. 2018, 2019b)	Finite element (FE) method	Abaqus	Deformation failure mode (out-of-plane displacement exceeding half the width of a layer). Non-linear geometric analysis considering Mohr–Coulomb elasto-plastic behavior and geometric imperfection	Perfect contact	Geometric imperfections considered using the profile of the first elastic buckling mode	Linear evolution of mechanical properties (Young's modulus and cohesion) as a function of time	Cylinder and wall
(Vantyghem et al. 2021)	FE method (automatic meshing using the voxel print method)	Grasshoper plugin by Rhinoceros and Abaqus	Possibility of modeling different modes of failure (elastic buckling, plastic collapse, deformation failure using non-linear geometric analysis)	Perfect contact	Geometric imperfections can be taken into account	Linear evolution of mechanical properties as a function of time	Simple and complex structures

(Ooms et al. 2021)	**FE method (automatic meshing using the cobra print method)**	**Grasshoper plugin by Rhinoceros and Abaqus**	**Possibility of modeling different modes of failure (elastic buckling, plastic collapse, deformation failure using nonlinear geometric analysis)**	**Interface model (tangential behavior characterized by the Coulomb friction model)**	**Geometric imperfections that can be added by randomly translating the positions of nodes in the original mesh**	**Linear evolution of mechanical properties as a function of time**	**Simple and complex structures**
(Liu et al. 2022)	FE method	Abaqus	Failure mode in deformation. Nonlinear geometric analysis considering Drucker–Prager elasto-plastic behavior and geometric imperfection	Perfect contact	Geometric imperfections considered using the profile of the first elastic buckling mode	Linear evolution of mechanical properties (Young's modulus and cohesion) as a function of time	Cylinder and wall
(Chang et al. 2021, 2022)	Lattice model (the designed object is discretized by a network of beam elements)		Deformation failure mode (horizontal displacement exceeding the width of a layer)	Perfect contact	Geometric nonlinearity added without the need to introduce initial geometric imperfections	Linear or exponential evolution of mechanical properties as a function of time	Cylinder, wall and square

Table 7.1. *Digital models of the 3D printing process at structural scale*

While initial studies have demonstrated the ability of numerical modeling to describe the 3D concrete printing process, a great deal of work has been carried out to improve the accuracy of predictions. For example, to simplify mesh creation for complex structures, various authors (Ooms et al. 2021; Vantyghem et al. 2021) propose the technique of automation using data transferred from CAD software in the geometric design phase to finite element analysis software. This approach removes all manual pre-processing steps from the digital modeling of the 3D printing process. Under this approach, the geometric model of the structure is imported and discretized automatically using the Grasshopper programming language, regardless of the complexity of the geometry to be printed. The file created contains all the information required (mesh, materials, boundary conditions, interactions and calculation steps) for numerical modeling in FE software (e.g. SIMULIA Abaqus/CAE). Technically, the discretization of the geometric model of the printed structure can be achieved in different ways. In the work by Vantyghem et al. (2021), the authors present a method called VoxelPrint, which involves voxelizing the geometric model into a set of 3D unit cubes (i.e. voxels). This technique uses the same process as for rasterizing a vector graphic into a raster image (a series of pixels). The main idea behind the voxelization algorithm is that a 3D space around the model is filled with many discrete cubes, and if the center of that cube lies within a predefined area of the 3D model and print path, the cube is considered to belong to the printed geometry. In this way, any 3D shape, no matter how complex, can be voxelized, and FE mesh generation is straightforward. One advantage of VoxelPrint is that a single large voxelized model is created (a single piece in the FE software), where interactions, binding constraints or contact properties are not specifically defined. The disadvantage of this method is that it does not allow the use of advanced contact properties, for example, to model joints, that is, weakened behavior between layers. In the calculation input file, mesh elements (voxels) are first deactivated in the initial step, using the model change function, and then progressively reactivated in each subsequent analysis step. This voxel-based simulation approach has been validated by comparison with the results of Wolfs et al. (2019b) in the case of a cylinder, where it provides a similar prediction of the maximum print height, after which the structure collapses (Vantyghem et al. 2021). The applicability of VoxelPrint was also demonstrated in Vantyghem et al. (2021) by its ability to distinguish failure by elastic buckling from failure by plastic deformation.

Another technique called CobraPrint allows the mesh to be generated by Grasshopper, first dividing the print path into multiple segments, and then assigning multiple mesh elements to each segment. As in the previous method, the segments are activated sequentially along the print path to simulate the printing process (Wolfs et al. 2018, 2019b; Ooms et al. 2021). In addition, the Coulomb friction model is used to present the tangential interaction between layers and between segments of each layer. Comparing with existing studies (straight wall (Wolfs et al. 2019b) and cylinder (Wolfs et al. 2018)), this technique showed an improvement in the prediction of fracture height without the need to incorporate initial geometric imperfections.

In order to simulate the whole 3D printing process by following the real printing path of complex structures, Nguyen-Van et al. (2021) proposed a new modeling procedure (virtual printing procedure). The principle of this procedure is to integrate the G-code files of the toolpath patterns into the simulation process. This procedure makes it possible to simulate the failure modes of structures during 3D printing, whatever their complexity. Examples include bio-inspired cellular blocks containing continuous non-self-intersecting surfaces, known as the triply periodic minimum surface (TPMS), gyroids and primitive blocks. The developed model was validated by comparison with experimental results on a printed hollow cylinder (Panda et al. 2019). The ability to predict both failure modes as well as the deformed shape of the layers was demonstrated. Not taking into account the effect of geometrical imperfections explains a moderate overestimation of 4.3% of the numerical prediction of the fracture height, compared to the experimental value.

At present, to model the plastic collapse of the concrete structure during 3D printing, the M-C perfect elasto-plastic model is widely adopted (Wolfs et al. 2018, 2019b). However, in a recent contribution, Liu et al. (2022) chose another model, the Drucker–Prager (D-P) model, to simulate the buildability of the printed structure according to the deformation failure mode. According to these authors, the D-P model provides a better representation of the stress–strain relationship, particularly in the post-peak phase, by using the strain-hardening function calibrated from the uniaxial stress-strain curve of early-age concrete. More specifically, using the same printed structures (e.g. the hollow cylinder (Wolfs et al. 2018) or the straight wall), replacing the M-C perfect elasto-plastic model with the D-P model shows an improvement, with the difference between numerical predictions and experimental results dropping from 58.6% to 44.8% in the cylinder case (Wolfs et al. 2018). In the case where elastic buckling is the dominant failure mode (straight wall case), the predicted results with the D-P and M-C models are identical as expected.

In parallel with finite element methods, Chang et al. (2020) have used the lattice model for numerical analysis of the 3D printing process. This method, based on a 3D network of Timoshenko beams, makes it possible to take into account the heterogeneous characteristics of printed objects and the influence of non-uniform gravitational loading. In this numerical model, structural failure according to the strain failure mode occurs when the lateral displacement exceeds the width of an individual layer, while material failure or plastic collapse results from the accumulation of local damage in the lower layers (Chang et al. 2021, 2022). The model correctly reproduced both failure modes, and the predicted critical print height improves on existing models in the literature (Wolfs et al. 2018, 2019b; Suiker et al. 2020). Comparing with experimental data in the cylinder case (Wolfs et al. 2018), the lattice model prediction overestimates the critical print height by 41.4%, which is lower than the 58.6% overestimation by the FE model of Wolfs et al. (2018). For the case of the square wall (Suiker et al. 2020), the relative difference between the two models is of the order of 10%. In the case where the structure fails by elastic buckling, as in the straight wall case (Wolfs et al. 2019b), the critical height predicted by the lattice model is around 12% lower than the experimental data. This lattice model has also demonstrated its ability to model unconfined uni-axial compression tests to calibrate the stiffness and strength of beam elements as a function of time.

7.3.2. *Simulation of the material deposition process at printed layer scale*

The material deposition phase is critical and requires high precision to ensure that the final product meets the design specifications and that the printed layers conform to the desired geometry. In the 3D concrete extrusion printing process, deposition strategies fall between two asymptotic regimes at the nozzle (Roussel 2018). In the first case, known as infinite brick extrusion, the extruded material is rigid and has a high initial yield strength, enabling it to resist gravity stresses and stresses due to pumping pressure. The material is therefore not sheared during extrusion, so that the filament cross-section is dictated by the shape of the nozzle. In the second case, known as "free-flow deposition", the material is sheared by a local contraction or screw mixer before it exits the nozzle. In this case, the final layer geometry results from a competition between gravity and the material's stress limit. It should be noted that in some cases, where the distance between the nozzle and the previous layer is less than the nozzle opening, the filament is shaped by being pressed between the nozzle and the substrate, until the filament thickness is equal to the distance between two successive nozzle passes. This strategy enables the layer thickness to be controlled, but it can generate local pressure on the lower layers, leading to further deformation of the layer and thus threatening the overall stability of the printed object.

Depending on the extrusion process chosen and the properties of the material, the final cross-section of a printed layer can take different forms, the exact prediction of which is a major issue for concrete 3D printing technology. The majority of concrete printers operate using cutting software, which assumes that the extruded filament will conform to a cross-section theoretically determined by applying mass conservation principles and using a filament thickness equal to the gap between the nozzle and the substrate. In reality, there is always a slight difference between this theoretical thickness and the actual thickness, which can accumulate over several layers and can subsequently lead to a significant change in the cross-sectional profile of the printed filaments. These deviations influence the aesthetic quality of the final product and can also lead to geometric non-conformities that can impact the structural properties (adhesion strength and durability) of the resulting printed product; and, in the worst case, to printing failures. In addition, depending on the kinematic process parameters chosen, the filament is susceptible to tearing, cracking or bending in the direction of deposition. The expected, well-controlled cross-section of the extruded filament is achieved when printing is carried out at nominal speed (i.e. when the material flow is equal to the product of the robot speed and the expected filament cross-section) (Perrot et al. 2021; Wolfs et al. 2021). However, in 3D printing installations, material flow at the nozzle is generally less precisely controlled than other process parameters, such as robot trajectory or speed.

Starting from the obvious importance of filament shape control and prediction, numerical studies of additive concrete manufacturing at the filament scale have been widely presented in a number of works. The so-called finite volume method (MVF), based on the computational fluid dynamics approach (CFD) is used in Comminal et al. (2019, 2020a, 2020b) to simulate concrete extrusion. In their study, Comminal et al. (2020b) analyze the effect of nozzle position height and the ratio between printing speed V and extrusion speed U on the geometry of the layer printed on a flat surface for a cement-based mortar mix. The simulations are based on two formulations of the constitutive law, the generalized Newtonian model and the fluid elasto-viscoplastic model. These virtual printing simulations provide numerical values of velocity, concrete stress and pressure at the nozzle outlet, as well as a visual representation of the cross-section of the printed layers. The study revealed an increase in pressure under the nozzle as the height separating the nozzle from the previously deposited layer decreased.

Previously, the effect of different constitutive laws, covering the wide range of material behaviors, including shear thinning and viscoplastic material phenomena, on a single printed layer was analyzed by Comminal et al. (2019). The printing process is simulated by moving a nozzle horizontally. The cross-section of printed filaments obtained with three different models of non-Newtonian fluids (power law, Herschel–Buckley and Bingham) appears thicker and less wide compared to a reference Newtonian fluid with constant viscosity, which itself produces filaments with a thickness less than the theoretical cross-section generally used by slicing software.

To further validate the CFD model developed, Spangenberg et al. (2021) carried out a numerical simulation of up to three layers using an elasto-viscoplastic constitutive law. The model was validated by experimental data on 3D-printed walls, with high accuracy. The study also quantifies the extrusion load imposed on the layer, which was found to exceed the material's yield strength in some regions of previously printed layers, leading to layer deformation/flow. The model could be used to optimize printing parameters, but the difficulty lies in its high computation time (several days are needed to run a simulation).

In Reinold et al. (2019), a numerical model based on the particle finite element method (PFEM) is used to simulate 3D concrete printing. The PFEM is an updated mesh-based approach in a Lagrangian framework. The flow behavior of freshly extruded concrete is based on a regularized viscoplastic Bingham model (Papanastasiou 1987). Its behavior in the elastic regime prior to failure can be modeled by this Bingham model using a large viscosity value. 2D numerical analyses with different process parameters and material properties were carried out to control layer deformations and enable targeted design with an understanding of the forces developing in the vicinity of the nozzle. These analyses reveal that nozzle height in relation to print speed and extrusion rate are the process parameters that govern the tension/extrusion forces under the nozzle. Variation in nozzle width did not result in any significant changes in maximum force values. However, higher stress concentrations were observed for smaller extrusion nozzle widths. The pressure generated under the extrusion nozzle during extrusion appears to be six times greater than that induced by a layer's own weight. Dynamic interactions during the printing of a second layer on an existing layer were analyzed on the basis of the quantity of material extruded. These were used to optimize the printing process parameters so as to prevent the formation of plastic deformations in the base layer. Thus, this type of multi-layer printing simulation can provide highly detailed information not only on the cross-sectional shape of the layers for predicting buildability, but also for quantifying the quality of surface roughness and providing a basis for predicting inter-layer adhesion.

In a recent study (Wolfs et al. 2021), 2D and 3D CFD simulations were carried out, using the finite volume method (MVF) to study the influence of the various parameters of the 3D printing process by extruding infinite bricks. These parameters include the printing speed V, the extrusion speed U, the geometric characteristics of the filament and nozzle (nozzle width and height between two layers) and the elasto-viscoplastic properties of the material. They showed that, when the process is run at nominal speed (V = U), the filament is extruded correctly and appears to be independent of the studied range of material properties and nozzle characteristics. When the extrusion speed is higher than the nozzle speed, U > V, the filament is observed to exhibit periodic elastic instabilities both in geometry (flexion behavior, referred to as filament buckling in Wolfs et al. (2021)) and stress distribution throughout extrusion. An increase in nozzle position height, a reduction in nozzle width or an increase in material consistency (i.e. greater stiffness, yield strength or viscosity) had a profound effect on the flexion of the filament. On the contrary, an extrusion speed lower than the nozzle speed, U < V, causes tensile stresses which can lead to cracking or even tearing of the filament. Similarly, an increase in material stiffness or a reduction in the material's critical strain increased the occurrence of tears. Analytical expressions for the tear factor and buckling factor were presented to indicate transitions from well-controlled nominal extrusion to filament tearing and buckling.

7.4. Discussions on recent advances, limitations and future research directions

The growing involvement of the academic and industrial sectors in the field of 3D concrete printing is gradually addressing the specific technical challenges associated with this innovative construction process. As previously presented, the numerical modeling tools developed have made it possible to detect and predict various failure modes throughout the printing process, whether they occur at the scale of a printed layer or of the entire structure. Numerical modeling has proved to be a valuable option for optimizing the 3D printing process, as it enables a systematic assessment of the influence of the entire range of materials, geometric parameters and process parameters.

Accurate prediction of structural behavior during printing relies on understanding each parameter involved in printing, such as material behavior and properties, geometric features and the printing environment. For example, the numerical study of the buckling of a 3D-printed structure by Wolfs et al. (2018), considering the time-dependent MC elasto-plastic behavior of the material, overestimated the printed heights before failure. The authors explain that compacting the material prior to the experimental tests (unconfined compression and direct shear tests) can lead to an overestimation of the material properties, which in turn can lead to an overestimation of the height before failure and an underestimation of the lateral deformations. Furthermore, in Wolfs and Suiker (2019), numerical predictions lose accuracy due to the influence of thermal heating of the 3D printer during the long printing process, which has not yet been incorporated into experimental characterization or numerical analyses.

Deviations between predictions and print tests can also be explained by various sources of uncertainty, such as uncertainties in material property measurements, imperfections or spatial variation in the geometry (e.g. width, thickness and position) of deposited layers, which are neglected in predictions. According to the results of experimental tests (such as the uni-axial compression test and the direct shear test) carried out by Wolfs et al. (2018), a non-negligible dispersion in the shear strength and Young's modulus of concrete at young age was noted, where a coefficient of variation (COV) of around 20% was indicated for each parameter. This uncertainty is linked to the heterogeneous nature of mortars and the transition state from fresh concrete (viscous fluid) to solid state, which is extremely difficult to accurately characterize. In addition to the uncertainty in material properties, Suiker et al. (2020) have also shown that the eccentricity of printed layers can significantly affect elastic buckling failure during printing.

Until now, the analysis of the behavior of the printed structure has been based on a deterministic problem using mean values of fresh concrete properties calibrated from laboratory tests. Variations (or uncertainties) in these properties are neglected, while a perfect rectangular cross-section of the layer is considered, ignoring the geometric imperfections that occur during deposition. Furthermore, in most cases, numerical tools are used to predict the behavior and especially the heights before failure of the printed structure, but they are not yet exploited for their capabilities and potentials, particularly in optimizing the 3D concrete printing process. The first attempt to take into account the uncertainty of fresh concrete properties in buildability analyses and 3D printing process design optimization can be found in Kruger et al. (2020). In this contribution, the authors used the probabilistic approach such as the first-order reliability method (FORM) to assess the impact of the large variation in material properties ignored by the classical deterministic design model. The optimal solution for layer thickness and printing speed was determined in order to minimize printing time. However, this latest study only considers the plastic failure mode, and the elastic buckling phenomenon is not addressed.

In this context, Diab's thesis (Diab 2023) was carried out within the Lamé laboratory, as part of the European Interreg CIRMAP project (CIrcular economy via customizable furniture with Recycled Materials for public Places). In this research work, a comprehensive reliability analysis procedure was developed with the aim of enhancing the predictive capability of numerical simulations and increasing their applicability in optimizing the 3D concrete printing process (Diab et al. 2023,

p. 202). More specifically, the analyses carried out by stochastic modeling make it possible to estimate the probability of different failure modes of concrete structures during the printing process, taking into account various uncertainties, such as those of the mechanical properties of fresh concrete. In the proposed procedure, uncertainties must first be quantified. For example, in the contribution by these present authors (Diab et al. 2022), the probabilistic inversion technique based on Bayesian theorem was adopted to quantify uncertainties in the elasto-plastic behavior of fresh concrete. Based on the uniaxial compression and shear test results of Wolfs et al. (2018, p. 20), Bayesian inference is used to determine the a posteriori statistical distribution of five parameters that characterize the structure's two failure modes during printing (i.e. plastic failure and elastic buckling), as illustrated in Figure 7.1.

The results of the uncertainty quantification step are then used as input data for stochastic modeling of the printed structures. While the Monte Carlo simulation (MCS) approach is widely known as the benchmark for reliability analysis, it nevertheless requires a very large number of direct evaluations of the structure's response, which is numerically very costly. To reduce the considerable time required for these stochastic analyses, the MCS can be combined with a model reduction technique developed on the basis of machine learning. Recently, in contributions by the present authors (Diab et al. 2022, 2023), the combination of the Kriging meta-modeling technique with MCS has been adapted to the field of 3D concrete printing. The main idea of this technique, named AK-MCS, is to iteratively build the response surface of the structure on a very reasonable number of samples of random parameters (i.e. the five mechanical property parameters of fresh concrete). In each iteration, the estimation of the probability of failure of the structure by the MCS can be easily achieved by interpolating the results of the constructed metamodel. After each iteration, a new, more appropriate sample of random parameters is added to reconstruct the metamodel. This AK-MCS stochastic modeling procedure continues until the estimated probability converges. As an example, Figure 7.1 shows the results of deterministic modeling and stochastic analysis of a straight concrete wall during printing. The availability of the analytical solution in this case of a simple structure (Suiker 2018) validates, on the one hand, the results of deterministic modeling by FE and, on the other hand, the applicability of the MCS performed on a million samples verifies the effectiveness of the AK-MCS. The same results for the estimated probabilities of two failure modes (i.e. plastic failure, elastic buckling) were obtained by both methods, but stochastic modeling by AK-MCS requires fewer than a hundred direct assessments of the printed structure.

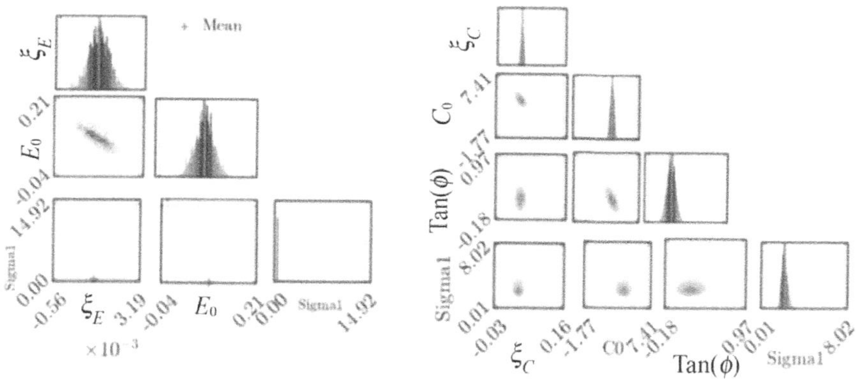

Parameters	E_0 (Pa)	ξ_E (Pa/min)	C_0 (Pa)	ξ_C (Pa/min)	Tan (ϕ)
Mean (μx)	79000	1200	3700	35	0.17
Standard deviation σx	26000	510	320	4.9	0.07
Variation coefficient (%)	32.91	42.5	8.65	14	41.2

Figure 7.1. *Quantifying uncertainty in the elasto-plastic properties of fresh concrete using Bayesian inference (Diab et al. 2022). For a color version of this figure, see www.iste.co.uk/perrot/3dconcrete.zip*

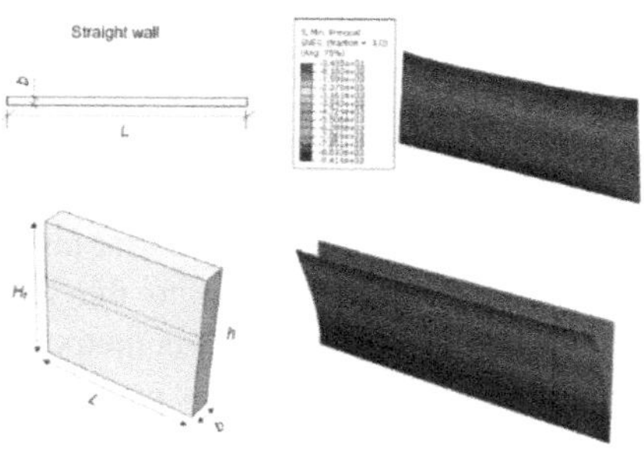

Method	Failure by elastic buckling	
	N_{call}	$P_f(\%)$
Direct MCS	10^6	49.10
AK-MCS	30	49.22
Method	**Failure by plastic collapse**	
	N_{call}	$P_f(\%)$
Direct MCS	10^6	40.68
AK-MCS	83	40.81

Figure 7.2. *Result of deterministic and reliability analyses of two failure modes (plastic failure, elastic buckling) of a straight wall during 3D printing and comparison of stochastic modeling by MCS and AK-MCS (Diab et al. 2023). For a color version of this figure, see www.iste.co.uk/perrot/3dconcrete.zip*

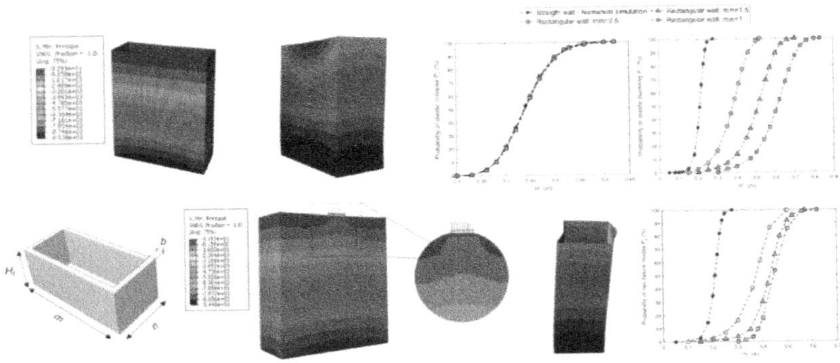

Figure 7.3. *AK-MCS reliability analyses of two failure modes (plastic failure, elastic buckling) of rectangular walls during 3D printing (Diab et al. 2023). For a color version of this figure, see www.iste.co.uk/perrot/3dconcrete.zip*

Reliability analyses of complex structures using the Kriging metamodel have been carried out in Diab et al. (2022, 2023). For example, in Diab et al. (2023), a numerical investigation with rectangular section walls verifies the geometrical effect on the risk of structural failure during printing. Figure 7.3 shows that the probability of plastic collapse of the square wall (m/n = 1) can be significantly higher than that of elastic buckling at a certain printing height. Conversely, for the rectangular wall with a large horizontal length ratio (e.g. m/n = 2.5), the probability of elastic buckling is much higher than that of plastic failure. Elastic buckling has also been identified as the dominant mode of failure in the case of a chair (Diab et al. 2022). Parametric studies have shown that the probability of elastic buckling depends strongly on the layer width of the printed chair, while the effect of printing speed and layer thickness is moderate (Figure 7.14).

The effect of the printing strategy, such as the layer pressing technique, is also taken into account in the reliability analyses. In addition to uncertainty in the mechanical properties of fresh concrete, probabilistic predictions of structural failure of rectangular walls consider the propagation of uncertainty related to the layer pressing strategy (e.g. uncertainty in extrusion speed, effective layer geometry and pressing force). Nevertheless, within the adopted range of random input parameters, it appears that the effect of the layer pressing strategy is less than that of the mechanical properties of the fresh concrete. Quantitatively, the global sensitivity analysis using Sobol indices identifies the parameters that most affect the probability of the two failure modes. The results in Figure 7.5 show, in the case studied, a strong influence of material properties, notably initial Young's modulus on the probability of elastic buckling, or initial cohesion and friction angle on the probability of plastic failure of printed structures.

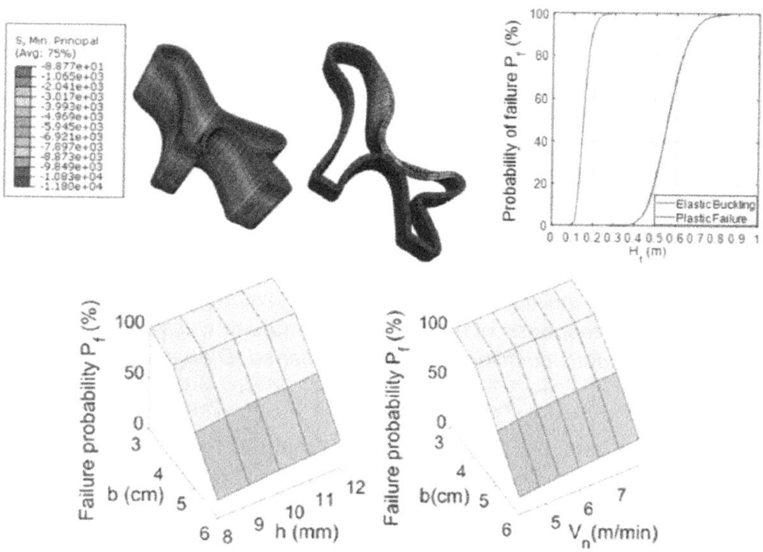

Figure 7.4. *AK-MCS reliability analysis of a chair made by 3D printing and variation of the probability of elastic buckling as a function of the geometry (e.g. thickness and width) of the printed layer and the printing speed (Diab et al. 2022). For a color version of this figure, see www.iste.co.uk/perrot/3dconcrete.zip*

The final stage of the methodology developed in Diab's thesis (Diab 2023) involves combining the AK-MCS technique with an optimization approach for the 3D printing process. In current developments, the chosen technique, named Kriging-QMC (Moustapha et al. 2016; Do et al. 2021), is based on quantile calculations of each failure mode and aims to optimize design parameters such as printing speed and deposited layer width by minimizing the amount of concrete and printing time. The first application in the case of the straight wall shows a good match between this Kriging-QMC technique and the well-known two-level optimization method using the MCS, which is, however, numerically very costly. Validation of this Kriging-QMC design optimization technique provides an effective tool for optimizing the 3D printing process for more complex concrete structures. An attempt to calibrate safety factors from these results to facilitate engineering applications is also envisaged.

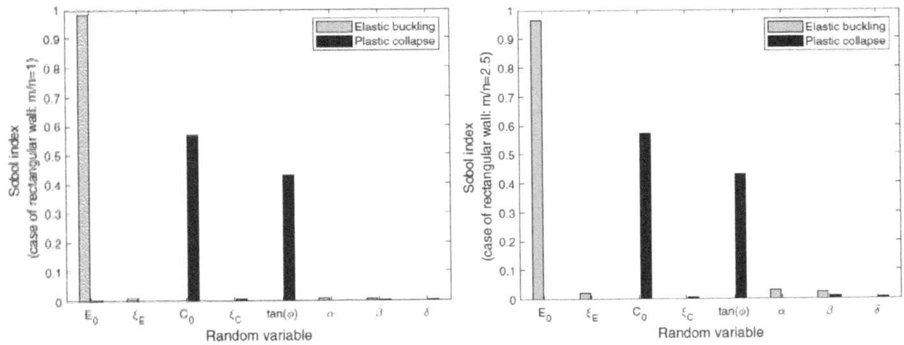

Figure 7.5. *Sobol index of input parameters provided by global sensitivity analysis on the response of rectangular and square walls during printing against elastic buckling and plastic collapse (Diab et al. 2023)*

It is important to point out that different assumptions have been adopted in these reliability analysis and design optimization studies using stochastic modeling. On the one hand, the variation in the mechanical properties of the fresh concrete in each layer during printing has been ignored to facilitate numerical modeling. This means that each deposited layer is considered homogeneous, and the mechanical properties of the printed structure vary only in the direction of extrusion. A more rigorous study of the variation in mechanical properties within each deposited layer (Nguyen-Van et al. 2021; Ooms et al. 2021; Vantyghem et al. 2021) would increase the accuracy of predictions. On the other hand, in our FE numerical studies, shell-type elements were chosen instead of volume elements, given the small thickness of the concrete structure compared to other dimensions. Although the 3D model using volumetric elements can provide the most accurate results of the structure's response, the simplified model based on surface elements enables us to reduce calculation time enormously, particularly in the context of probabilistic analyses and design optimization using stochastic modeling. It should also be noted that, in these current studies, the geometric imperfection and possible spatial heterogeneity of the geometry (e.g. thickness and width) of printed layers is not taken into account, and only the analysis of bifurcation buckling has been undertaken.

The developments and applications of numerical tools in the 3D printing process for concrete structures as previously presented have demonstrated the effectiveness of numerical modeling in this field. Its high performance, particularly in predicting the behavior of complex structures in the design phase and during printing, offers great advantages over studies based on analytical solutions. However, a major limitation of this numerical approach is the considerable increase in computational time and resources, depending on the volume of the large-scale structure, the desired accuracy and the target objective. This limitation largely explains why numerical studies have been focused on each specific scale (e.g. deposited layer scale, structure scale) and why, to date, there is no universal model or tool that can predict the entire 3D printing process. The challenge for the development of such a tool is further increased by taking into account physical phenomena (e.g. the influence of temperature, relative humidity, etc.), which have been very rarely discussed to date.

Despite its constant development, the implementation of 3D concrete printing in real construction projects remains complex. Indeed, the engineering principles governing construction, as in the building sector, are highly standardized and regulated, and relate as much to the materials used as to the amount of structural reinforcement and the methodology of the construction technique. Compliance with these specific regional standards and codes is essential to guarantee the safety and integrity of the structure in accordance with its design, location and function. However, as a new technology, 3D concrete printing does not yet have standardized construction rules to establish a clear framework for its use in construction projects. The concrete 3D printing community presents the first attempt to create norms and standards (Buswell et al. 2020) to broaden the technology's adoption in the industry. In addition, 3D printing companies have found ways to overcome the certification hurdle in order to bring their projects to fruition. For example, a bridge was produced using the Eindhoven University of Technology printer (TU/e) and certified for public use after validation by testing on a scale model (Salet et al. 2018). The current procedure is based on extensive testing of materials and the final structure to provide a satisfactory guarantee of safety and compliance with traditional codes. However, this strategy requires much more material, contradicting the aim of this new technology to reduce material consumption during construction and limiting the development and application of 3D concrete printing on a large scale in practice. Similarly, in 2016, taking advantage of the architectural freedom offered by 3D concrete printing, XtreeE 3D printed a pillar of complex geometry for a public building, which could not have been easily manufactured with any other technology (Gaudillière et al. 2020). However, in the absence of specific regulations, only the outer wall of the structure was 3D printed, serving as a mold for the concrete cast inside. The result was an oversized final structure, since only the material poured inside was taken into account in the structural analysis, and the printed material was considered to play no mechanical role. These cases underline the need for regulations and an optimization process to ensure proper design.

The development and optimization of numerical simulation tools and procedures for industrial-scale applications is therefore a major objective for 3D printing.

While maintaining the necessary accuracy of predictions, these numerical tools need to be able to deliver results continuously and almost instantaneously. This is an essential prerequisite for decision support and real-time optimization of the 3D printing process. The stochastic optimization procedure could be widely useful for establishing design regulations for structures built using 3D printing. To approach this objective, a number of options could be considered:

– Results from numerical modeling at the CFD layer scale could be better exploited to characterize the interaction between different printing parameters (e.g. extrusion speed, deposited layer width and thickness and nozzle geometry). These numerical tools can also be used to quantify the uncertainty of these printing parameters and of the force imposed by the layer pressing technique, which are not yet accessible.

– Consideration of geometric imperfections and spatial heterogeneity in the mechanical properties and self-weight of printed layers in probabilistic modeling would improve predictions at the structural scale. However, the necessary use of volume elements, which is numerically very costly, will be an obstacle to overcome in these simulations. Enhancing the capability of numerical tools by combining them with model reduction techniques using machine learning algorithms would be essential in this context.

From a strategic point of view, using the approach based on direct assessments of the structure's behavior seems to be very difficult, if not inapplicable, for real-time optimization of the 3D printing process. While this approach could be very useful in the design phase prior to manufacturing, another approach should be adopted to study and optimize the structure during the printing phase. Artificial intelligence (AI), using deep learning tools, is a very promising approach. Thanks to this method, it would be possible to characterize and quantify (in real time) the various parameters (e.g. mechanical properties of fresh concrete, nozzle pressure on already printed layers, etc.) using continuous measurement data during printing (e.g. measurements of vertical displacement or deformation of the structure, etc.). These results would provide important data for the optimization of structure printing, which would be carried out in a continuous and sequential manner. More precisely, corresponding to a given iteration or point in time during the printing phase, updating the input parameters and their associated uncertainties would help verify the predictions and decisions of the previous iteration. These would then be taken into account to predict the behavior and optimize the printing of the structure in the next iteration. The effectiveness of such an approach would closely depend on the performance of the deep learning technique and also on the quality and quantity of the data gathered from the printing tests, which are also supplemented by the results of prior numerical predictions.

7.5. Conclusion

Digital tools play an essential role in the three-dimensional printing process for concrete structures. From the design phase of the geometric model in the virtual environment to the construction of the actual structure and finishing for commissioning, these tools are becoming an essential choice for this new construction technique. The recent development of digital tools makes it possible, on the one hand, to automate all the various stages of the 3D printing process (design, layering of the virtual geometric model, definition of printer trajectories and speeds); and, on the other hand, to directly supply input files for the digital modeling of components and/or the printed structure.

This chapter summarizes the state of the art in numerical modeling of the 3D printing process, focusing on the scale of the structure and the printed layer. On a large scale, the interest of numerical modeling lies in predicting failure modes such as elastic buckling and plastic collapse of the structure, phenomena commonly observed during printing. The results obtained at this scale demonstrate the performance of numerical studies based largely on the finite element method, adopting the material's elasto-plastic behavior model. At the filament scale, numerical modeling involves predicting the shape of the deposited layers. In all studies at this level, a viscous behavior represented by a Newtonian or non-Newtonian fluid model is chosen to characterize the state of fresh concrete, while simulations can be carried out using finite volume or finite element-particle methods. Numerical predictions have demonstrated their ability to simulate the cross-section of the printed layer, which depends not only on the parameters and strategy of the printing process but also on the properties of the material and its interaction with the existing layer. The ability to characterize the quality of the contact surface roughness between successive layers, or the pressure force under the nozzle, is also an advantage of numerical modeling at this scale.

However, numerical predictions sometimes differ significantly from experimental results. The difference can be explained by an overestimation of the mechanical properties of fresh concrete calibrated on the basis of experimental tests, or by the influence of thermal phenomena not yet taken into account in numerical simulations. Among the various ways of improving the predictive capacity of numerical modeling, the stochastic analysis approach offers a promising avenue. On the one hand, this type of reliability analysis approach makes it possible to quantify sources of uncertainty, such as the uncertainty of the mechanical properties of concrete at a young age, and their propagation to the responses of printed structures.

On the other hand, by combining them with the appropriate algorithms, stochastic modeling can be used to optimize the printing process corresponding to a chosen confidence level. Moreover, the ability to calibrate safety factors on the basis of these optimization results would contribute to the standardization of construction using the 3D printing process.

The future development and enhancement of digital tools must meet the need for real-time optimization and/or decision support of the printing process, while maintaining the necessary precision. The use of AI, in which data is collected from measurements during the printing process and from digital simulations at different scales in the design phase, would enable this objective to be approached.

7.6. References

Abdulhameed, O., Al-Ahmari, A., Ameen, W., Mian, S.H. (2019). Additive manufacturing: Challenges, trends, and applications. *Advances in Mechanical Engineering*, 11(2), 1–27.

Anane, W., Iordanova, I., Ouellet-Plamondon, C. (2021). The use of BIM for robotic 3D concrete printing. In *Canadian Society of Civil Engineering Annual Conference*. Springer, Singapore.

Buswell, R.A., Da Silva, W.L., Bos, F.P., Schipper, H.R., Lowke, D., Hack, N., Kloft, H., Mechtcherine, V., Wangler, T., Roussel, N. (2020). A process classification framework for defining and describing digital fabrication with concrete. *Cement and Concrete Research*, 134, 106068.

Chang, Z., Schlangen, E., Šavija, B. (2020). Extended lattice model to simulate the printing process of 3D printed cementitious materials. In *Second RILEM International Conference on Concrete and Digital Fabrication: Digital Concrete 2020*. Springer, Cham.

Chang, Z., Xu, Y., Chen, Y., Gan, Y., Schlangen, E., Šavija, B. (2021). A discrete lattice model for assessment of buildability performance of 3D-printed concrete. *Computer-Aided Civil and Infrastructure Engineering*, 36(5), 638–655.

Chang, Z., Zhang, H., Liang, M., Schlangen, E., Šavija, B. (2022). Numerical simulation of elastic buckling in 3D concrete printing using the lattice model with geometric nonlinearity. *Automation in Construction*, 142, 104485.

Comminal, R., Serdeczny, M.P., Ranjbar, N., Mehrali, M., Pedersen, D.B., Stang, H., Spangenberg, J. (2019). Modelling of material deposition in big area additive manufacturing and 3D concrete printing. In *Joint Special Interest Group Meeting between Euspen and ASPE Advancing Precision in Additive Manufacturing*. The European Society for Precision Engineering and Nanotechnology, Cranfield, Bedfordshire.

Comminal, R., da Silva, W.R.L., Andersen, T.J., Stang, H., Spangenberg, J. (2020a). Influence of processing parameters on the layer geometry in 3D concrete printing: Experiments and modelling. In *RILEM International Conference on Concrete and Digital Fabrication*. Springer, Cham.

Comminal, R., da Silva, W.R.L., Andersen, T.J., Stang, H., Spangenberg, J. (2020b). Modelling of 3D concrete printing based on computational fluid dynamics. *Cement and Concrete Research*, 138, 106256.

Diab, Z. (2023). Application d'outils numériques à l'analyse probabiliste de la constructibilité et à l'optimisation du procédé d'impression 3D. PhD Thesis, Université d'Orléans, Orléans.

Diab, Z., Do, D.P., Rémond, S., Hoxha, D. (2022). Uncertainty quantification of concrete properties at fresh state and stability analysis of the 3D printing process by stochastic approach. In *RILEM International Conference on Concrete and Digital Fabrication*. Springer, Cham.

Diab, Z., Do, D.-P., Rémond, S., Hoxha, D. (2023). Probabilistic prediction of structural failure during 3D concrete printing processes. *Materials and Structures*, 56(4), 73.

Ding, Z., Liu, S., Liao, L., Zhang, L. (2019). A digital construction framework integrating building information modeling and reverse engineering technologies for renovation projects. *Automation in Construction*, 102, 45–58.

Do, D.P., Tran, N.T., Hoxha, D., Vu, M.N., Armand, G. (2021). Kriging-based optimization design of deep tunnel in the rheological Burger rock. In *IOP Conference Series: Earth and Environmental Science*, Turin, 833, 012155.

Eastman, C.M. (2011). *BIM Handbook: A Guide to Building Information Modeling for Owners, Managers, Designers, Engineers and Contractors*. John Wiley & Sons, New York.

Fologram for Rhino and Grasshopper (n.d.). [Online]. Available at: https://www.food4rhino.com/en/.

Gaudillière, N., Dirrenberger, J., Duballet, R., Bouyssou, C., Mallet, A., Roux, P., Zakeri, M., Xtree, E.R. (2020). Industrialising concrete 3D printing: Three case studies. In *Design Transactions: Rethinking Information Modelling for a New Material Age*, Sheil, B., Thomsen, M.R., Tamke, M., Hanna, S. (eds). UCL Press, London.

Keita, E., Bessaies-Bey, H., Zuo, W., Belin, P., Roussel, N. (2019). Weak bond strength between successive layers in extrusion-based additive manufacturing: Measurement and physical origin. *Cement and Concrete Research*, 123, 105787.

Kruger, J., Zeranka, S., van Zijl, G. (2019). 3D concrete printing: A lower bound analytical model for buildability performance quantification. *Automation in Construction*, 106, 102904.

Kruger, J., Cho, S., Zeranka, S., Viljoen, C., van Zijl, G. (2020). 3D concrete printer parameter optimisation for high rate digital construction avoiding plastic collapse. *Composites Part B: Engineering*, 183, 107660.

Le, T.T., Austin, S.A., Lim, S., Buswell, R.A., Law, R., Gibb, A.G., Thorpe, T. (2012). Hardened properties of high-performance printing concrete. *Cement and Concrete Research*, 42(3), 558–566.

Lim, S., Buswell, R.A., Valentine, P.J., Piker, D., Austin, S.A., De Kestelier, X. (2016). Modelling curved-layered printing paths for fabricating large-scale construction components. *Additive Manufacturing*, 12, 216–230.

Liu, H., Ding, T., Xiao, J., Mechtcherine, V. (2022). Buildability prediction of 3D–printed concrete at early-ages: A numerical study with Drucker–Prager model. *Additive Manufacturing*, 55, 102821.

McNeel, R. (2010). Rhinoceros 3D, Version 6.0. Robert McNeel & Associates, Seattle, WA.

Moustapha, M., Sudret, B., Bourinet, J.-M., Guillaume, B. (2016). Quantile-based optimization under uncertainties using adaptive Kriging surrogate models. *Structural and Multidisciplinary Optimization*, 54, 1403–1421.

Nerella, V.N., Hempel, S., Mechtcherine, V. (2019). Effects of layer-interface properties on mechanical performance of concrete elements produced by extrusion-based 3D-printing. *Construction and Building Materials*, 205, 586–601.

Nguyen-Van, V., Panda, B., Zhang, G., Nguyen-Xuan, H., Tran, P. (2021). Digital design computing and modelling for 3-D concrete printing. *Automation in Construction*, 123, 103529.

Ooms, T., Vantyghem, G., Van Coile, R., De Corte, W. (2021). A parametric modelling strategy for the numerical simulation of 3D concrete printing with complex geometries. *Additive Manufacturing*, 38, 101743.

Panda, B., Lim, J.H., Tan, M.J. (2019). Mechanical properties and deformation behaviour of early age concrete in the context of digital construction. *Composites Part B: Engineering*, 165, 563–571.

Papanastasiou, T.C. (1987). Flows of materials with yield. *Journal of Rheology*, 31(5), 385–404.

Perrot, A., Rangeard, D., Pierre, A. (2016). Structural built-up of cement-based materials used for 3D-printing extrusion techniques. *Materials and Structures*, 49, 1213–1220.

Perrot, A., Pierre, A., Nerella, V.N., Wolfs, R.J.M., Keita, E., Nair, S.A.O., Neithalath, N., Roussel, N., Mechtcherine, V. (2021). From analytical methods to numerical simulations: A process engineering toolbox for 3D concrete printing. *Cement and Concrete Composites*, 122, 104164.

Reinold, J., Nerella, V.N., Mechtcherine, V., Meschke, G. (2019). Particle finite element simulation of extrusion processes of fresh concrete during 3d-concrete-printing. In *Sim-AM 2019: Second International Conference on Simulation for Additive Manufacturing*. CIMNE, Barcelona.

Reiter, L., Wangler, T., Roussel, N., Flatt, R.J. (2018). The role of early age structural build-up in digital fabrication with concrete. *Cement and Concrete Research*, 112, 86–95.

Roussel, N. (2018). Rheological requirements for printable concretes. *Cement and Concrete Research*, 112, 76–85.

Roussel, N., Spangenberg, J., Wallevik, J., Wolfs, R. (2020). Numerical simulations of concrete processing: From standard formative casting to additive manufacturing. *Cement and Concrete Research*, 135, 106075.

Salet, T.A., Ahmed, Z.Y., Bos, F.P., Laagland, H.L. (2018). Design of a 3D printed concrete bridge by testing. *Virtual and Physical Prototyping*, 13(3), 222–236.

Spangenberg, J., da Silva, W.R.L., Comminal, R., Mollah, M.T., Andersen, T.J., Stang, H. (2021). Numerical simulation of multi-layer 3D concrete printing. *RILEM Technical Letters*, 6, 119–123.

Suiker, A.S.J. (2018). Mechanical performance of wall structures in 3D printing processes: Theory, design tools and experiments. *International Journal of Mechanical Sciences*, 137, 145–170.

Suiker, A.S.J., Wolfs, R.J., Lucas, S.M., Salet, T.A. (2020). Elastic buckling and plastic collapse during 3D concrete printing. *Cement and Concrete Research*, 135, 106016.

Tay, Y.W.D., Panda, B., Paul, S.C., Noor Mohamed, N.A., Tan, M.J., Leong, K.F. (2017). 3D printing trends in building and construction industry: A review. *Virtual and Physical Prototyping*, 12(3), 261–276.

Teizer, J., Blickle, A., King, T., Leitzbach, O., Guenther, D., Mattern, H., König, M. (2018). BIM for 3D printing in construction. *Building Information Modeling: Technology Foundations and Industry Practice*, 421–446.

Vandezande, J., Krygiel, E., Read, P. (2012). *Autodesk Revit Architecture 2013: Essentials*. John Wiley & Sons, New York.

Vantyghem, G., Ooms, T., De Corte, W. (2021). VoxelPrint: A grasshopper plug-in for voxel-based numerical simulation of concrete printing. *Automation in Construction*, 122, 103469.

Wangler, T., Lloret, E., Reiter, L., Hack, N., Gramazio, F., Kohler, M., Bernhard, M., Dillenburger, B., Buchli, J., Roussel, N. (2016). Digital concrete: Opportunities and challenges. *RILEM Technical Letters*, 1, 67–75.

Wolfs, R.R. (2015). 3D printing of concrete structures. Master's Thesis, Eindhoven University of Technology.

Wolfs, R.J.M. and Suiker, A.S.J. (2019). Structural failure during extrusion-based 3D printing processes. *The International Journal of Advanced Manufacturing Technology*, 104, 565–584.

Wolfs, R.J.M., Bos, F.P., Salet, T.A.M. (2018). Early age mechanical behaviour of 3D printed concrete: Numerical modelling and experimental testing. *Cement and Concrete Research*, 106, 103–116.

Wolfs, R.J.M., Bos, F.P., Salet, T.A.M. (2019a). Hardened properties of 3D printed concrete: The influence of process parameters on interlayer adhesion. *Cement and Concrete Research*, 119, 132–140.

Wolfs, R.J.M., Bos, F.P., Salet, T.A.M. (2019b). Triaxial compression testing on early age concrete for numerical analysis of 3D concrete printing. *Cement and Concrete Composites*, 104, 103344.

Wolfs, R.J., Salet, T.A., Roussel, N. (2021). Filament geometry control in extrusion-based additive manufacturing of concrete: The good, the bad and the ugly. *Cement and Concrete Research*, 150, 106615.

Wu, P., Wang, J., Wang, X. (2016). A critical review of the use of 3D printing in the construction industry. *Automation in Construction*, 68, 21–31.

8

Environmental Impact of 3D Concrete Printing

Kateryna KUZMENKO[1]**, Charlotte ROUX**[2] **and Adélaïde FERRAILLE**[3]

[1] *Groupe Kardham, Paris, France*

[2] *École des Mines de Paris, France*

[3] *Laboratoire Navier, École des Ponts, Marne-la-Vallée, France*

Summary

The current development of 3D concrete printing technology is driven mainly by its potential to increase the productivity and formal freedom of concrete construction, as well as to reduce the industry's costs and environmental impact. The latter, based on the ability to reduce the material content of construction elements through rational material deposition, is studied in this work using the life cycle assessment (LCA) method.

The environmental model established here proposes a generic assessment framework that can be applied to all extrusion printing processes, whatever the material used. A specific separation between the impact of the material and that of the process is maintained throughout the environmental assessment carried out at the scale of the process and then of the constructive system, the space lattice wall system. The latter is then compared with its conventional analogues in the masonry and reinforced concrete categories. This shows that, despite the substantial material savings possible with the 3D printing process, environmental impact savings are not guaranteed at the scale of the constructive system, notably due to the impact linked to the process. We thus introduce the *Kf* coefficient, capturing the level of impact transfer from the material to the process for a given technology, and thus designing the most environmentally relevant industrial application.

3D Concrete Printing,
coordinated by Arnaud PERROT and Yohan JACQUET.

8.1. Introduction

Concrete is the most widely used building material in the world, ahead of all other materials (stone, clay, metal and wood) (Van Damme 2018). This intense use of concrete is mainly due to the wide availability of its components, their relatively low cost and the ease with which they can be transformed into a strong, durable, structural material.

Concrete is a mixture of water, sand and gravel, joined by a binder, usually cement. It is this last component, cement, which accounts for most of concrete's environmental profile (Gartner 2004). Indeed, the preparation of 1 ton of cement emits almost 1 ton of CO_2 equivalent (kg CO_2 eq). Half of these emissions come from the chemical reactions involved in the decarbonation of limestone, the other half from the energy (mostly of fossil origin) required for firing at 1,450°C, the temperature needed for decarbonation (Feraille 2016). Worldwide, cement production accounts for between 5% (Gartner 2004) and 8% (Flatt et al. 2012) of anthropogenic CO_2 emissions. The other constituents of concrete (gravel and sand), on the other hand, have a very low carbon footprint, even close to zero: 1 ton of aggregate corresponds to around 2.75 kg CO_2 eq (Union Nationale des Producteurs de Granulats 2017). Thus, the carbon footprint of 1 m^3 of concrete is approximately equal to the mass of the cement in the mix, that is, an average of 300 kg, equivalent to 300 CO_2 eq /m^3 (Kuzmenko 2021; Kuzmenko et al. 2022).

On the scale of the construction system, numerous studies have shown that the share of traditional placement processes (by pouring) represents between 1% and 2% of the total balance (Hong et al. 2015; Li et al. 2019; Ryklová et al. 2019; Kuzmenko 2021). Thus, the environmental impact of a concrete construction element will also be defined, for the most part, by the quantity of cement in the mixture.

There are two possible strategies for reducing the environmental impact of concrete construction. The first involves reducing the impact of the material itself, by substituting cement with alternative binders such as pozzolanic materials, limestone fillers or industrial residues (blast furnace slag and fly ash) (Gartner 2004; Flatt et al. 2012; Feraille 2016; Scrivener et al. 2018). The second is aimed at reducing the carbon footprint on the scale of a construction element/system, or even the entire industry, and thus proposes to reduce the overall volume of concrete use. This can be achieved by resorting to material reduction strategies and more virtuous material deposition methods, such as additive manufacturing technology aka 3D concrete printing (Gosselin et al. 2016; Wangler et al. 2016, 2019; Buswell et al. 2018; Duballet et al. 2019).

In this chapter, we will study the environmental relevance of this second strategy using the LCA method, and in particular the model developed in our previous work (Kuzmenko 2021; Kuzmenko et al. 2022; Roux et al. 2022). This LCA model for 3D concrete printing technology distinguishes between the share of impact related to the materials (concrete) and that related to the implementation technology (3D printing cell), and will be applied at two scales: the manufacturing process and the construction system. To study the latter, the case study of a space lattice wall construction system developed specifically for 3D concrete printing technology was considered. The environmental impact of the printed wall is then compared with its conventional analogues, thus outlining the industrial perspective of the practice of reducing the environmental impact of concrete construction.

8.2. 3D printing technology and case studies

8.2.1. *3D extrusion–deposition printing*

Large-scale additive manufacturing with cementitious materials brings together different types of processes. The founding concept was proposed by Pegna in the late 1990s (Pegna 1997) and was based on a "powder bed" technology activated by water and additives. However, on a large scale, the first experiments were carried out with another system, contour crafting (Khoshnevis 2004), giving rise to a set of processes later developed and grouped under the notion of digital concrete (Wangler et al. 2016, 2019). In 2022, the classification of different additive manufacturing processes was proposed by Buswell et al. (2020) in order to drive standardization and the creation of industry standards for the practice.

Here, we take a closer look at a technology from the contour crafting family known as "extrusion-deposition" (Perrot et al. 2018), initially developed in England at Loughborough University and currently being industrialized by the XtreeE company, as shown in Figure 8.1.

The printing cell consists of a six-degree-of-freedom robotic arm, equipped with a printing head that is fed with concrete from a concrete mixing, stirring and pumping unit. The whole system is digitally controlled by a computer from a 3D model that integrates information on the print path.

The extrusion–deposition process involves the successive stacking of layers of continuous concrete filaments, with cross-sections ranging from a few millimeters to a few centimeters. Provided that the mechanical properties of the concrete filaments evolve gradually, this process makes it possible to mass-produce non-standard, free-form or even complex elements, unlike the technique of casting concrete using formwork.

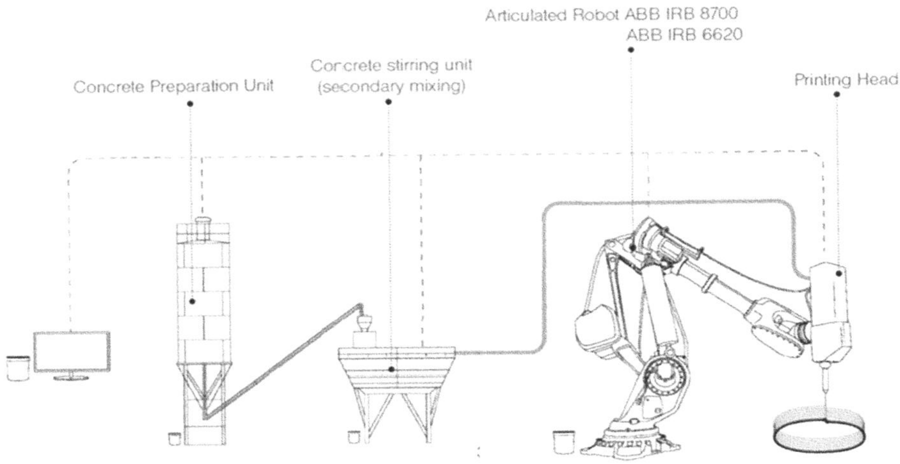

Figure 8.1. *The extrusion–deposition 3D printing cell. Image: Courtesy of XtreeE/courtesy of XtreeE*

8.2.2. *Material-saving construction elements*

The pioneers of this new process quickly transformed this possibility of producing "free" shapes (i.e. not constrained by the formwork shape) into a trend for reducing material within constructive elements. In particular, they resorted to topological optimization, that is, the removal of all structurally unnecessary material, thanks to the ability of additive manufacturing to deposit material freely.

a) b)

Figure 8.2. *Concrete 3D printing construction systems: (a, left) prestressed bicycle bridge (Salet et al. 2018); (b, right) space lattice wall system (Duballet 2019). For a color version of this figure, see www.iste.co.uk/perrot/3dconcrete.zip*

Figure 8.2 illustrates four examples of topologically optimized building elements produced using 3D concrete printing. The space lattice wall (Figure 8.2(a)), for example, reduces its weight per unit area by almost 80% compared with its conventional analogues (Duballet 2019; Duballet et al. 2019) and the funicular floor (Figure 8.2(c)) by around 70% (Rippmann et al. 2018).

8.2.3. *Case study*

It is therefore natural to associate environmental benefits with a reduction in the quantity of structural material for the same function. This hypothesis will be studied here in the case of the space lattice wall and presented below.

8.2.3.1. *Space lattice construction system*

The design, manufacture and rationale behind the space lattice wall system were originally described by Duballet (2019) and Duballet et al. (2019). In a nutshell, the structure was specifically designed for 3D concrete printing technology, taking advantage in particular of the benefit of "putting the material exactly where it's needed". In concrete terms, the system saves over 80% of material in terms of mass, compared with its conventional masonry analogues, while ensuring the required structural and thermal requirements (Duballet et al. 2019).

In the present study, the system is not considered as a specific structural solution, but rather as a generalization of topologically optimized structures produced with 3D concrete printing, enabling significant material savings thanks to the intelligent and rational deposition of the latter. In other words, we aim to assess the extent to which the material-saving approach leads to a real reduction in environmental impact, based on a specific case study representing this approach.

8.2.3.2. *Design and construction*

The system's morphology is based on a spatial tiling principle, subdividing a given volume into polyhedra. Concrete is printed on the edges of the polyhedra, shaping the latticework (see Figure 8.3). By filling the polyhedral volumes with a heat-resistant material, that is, polystyrene or cementitious foam, an insulating function can be achieved. Beyond vertical wall applications, the system as such can be extrapolated to almost any flat or curved membrane, that is, slab or roofing system. The present study will focus on a specific application of the system, that is, an insulating load-bearing wall, suitable for the exterior walls of building envelopes.

Figure 8.3. *(a, top) Prototype of the space lattice wall (Kuzmenko 2021). (b, bottom) Prototype of space lattice wall reinforced with continuous long fiber (Kuzmenko 2021). For a color version of this figure, see www.iste.co.uk/perrot/3dconcrete.zip*

The lattice wall construction process consists of the following steps: first, the insulating blocks are shaped and then progressively assembled to support the concrete extrusion. Once assembled, the blocks support and protect the concrete as it gradually acquires mechanical strength, preparing to support its own weight as well as the additional material above.

8.2.3.3. *Industrial implementation*

Industrial registration

The industrial implementation of 3D-printed construction systems is the subject of much debate in the field, with no universal consensus established. The key question is whether the printed structure should be considered as masonry (Eurocode 6) or as reinforced concrete (Eurocode 2).

Initially, the practice's founder, J. Pegna, presented his invention as a "new approach to masonry", rather than the generalization of concrete structures naturally implied by the material analogy. Later, the conceptual relevance of 3D concrete printing within the theoretical framework of masonry was underlined by a series of academic works (Duballet 2019, 2020; Carneau et al. 2020). With regard to the reinforced concrete framework, the current technological obstacle preventing the practice from operating within it, mainly concerns the integration of reinforcement in printed concrete. Several reinforcement strategies are currently being developed in academic and industrial circles (Scrivener et al. 2018; Ryklová et al. 2019; Wangler et al. 2019).

The space lattice wall initially has no reinforcement, being conceived as a masonry principle. From a normative point of view, this industrial framework would thus be governed by Eurocode 6 and would therefore be limited to applications such as a wall in a single-family home or a shear wall in a multi-storey building.

One reinforcement solution studied here was developed by Ducoulombier et al. (2020) and Demont et al. (2021) and involves integrating a continuous fiber into the printing filament along the printing direction. Placed uni-directionally in the printed filament, the fiber provides increased tensile strength, proportional to the volume of integrated fibers, and ensures ductile behavior always in the direction of printing. The final material is therefore highly anisotropic and ductile.

It is therefore a reinforcement solution naturally adapted to the space lattice construction system, as it is based on structural behavior combining compression and tensile strength. In this way, the structural model, initially designed and dimensioned for the masonry construction system, is generalized to the reinforced concrete category. The surrounding framework frames the internal latticework, reorganizing the redistribution of forces. Figure 8.3(b) illustrates the fabricated structure.

A first prototype of the space lattice as a masonry concept was produced and presented at IASS 2018 (Kuzmenko et al. 2022), as shown in Figure 8.3(a). A curved version of the space lattice including continuous fiber reinforcement was exhibited at IASS 2019, as shown in Figure 8.3(b).

8.2.4. *Study issues*

Once again, it seems natural to associate environmental benefits with a reduction in the quantity of structural material for the same function. However, not only does the quantity of material not seem to be a totally appropriate metric for environmental reasoning, but the change in implementation process also introduces new unknowns into the system. Indeed, with a printing cross-section of 1 cm^2, the six-degree-of-freedom robotic arm – weighing between 1 and 5 tons – travels 10 km to deposit 1 m^3 of concrete, and throughout this journey, pumping, mixing and depositing operations will consume energy. The energy consumption involved in placing concrete by printing appears to be much higher than that of formwork, and so the question arises: is this likely to cancel out the environmental benefits obtained through material savings? To answer this question, we use the LCA method.

8.3. Methodology and case studies

8.3.1. *Life cycle assessment method*

Framed by international standards ISO 14 040 and 14 044, then in Europe and France by standard EN 15 804 (specific to the construction sector), the LCA method consists of tracing all material and energy flows in and out of the system, from the extraction of raw materials to end-of-life. These flows are then translated into environmental impacts according to several categories: climate change, toxicity, water eutrophication, soil acidification, etc. LCA is therefore a multi-stage, multi-criteria method for quantifying the environmental impact of a construction system or product.

8.3.1.1. *Impact and pollution transfer*

The phenomena of impact and pollution transfer are crucial methodological elements of the LCA method. Both often occur after technological changes in a given system, and make it possible to monitor and quantify the *transfer of impact* from one life cycle phase to another, as well as the *transfer of pollution* from one environmental indicator to another. However, both can only be observed if the multi-stage, multi-criteria aspect of the method is respected during analysis.

It should be noted that the focus here is on the impact of climate change. However, other indicators are also worth looking at to avoid replacing one problem with another. For example, the impact on the consumption of abiotic resources is also of great interest in the case of concrete (sand being the second most used resource in the world after water). At present, the resource depletion indicator considered in a life cycle analysis does not take account of this issue, as it is based on the fact that silica is an almost inexhaustible resource on a global scale. Work is in progress to develop this indicator[1]. Generally speaking, in an LCA, a coherent set of indicators covering three main areas of protection (health, biodiversity and resources) is generally evaluated to obtain a complete picture of the environmental impact of the product or system under consideration.

In the present study, we will focus mainly on the phenomenon of *impact transfer* from one phase of the life cycle to another through the *Kf* coefficient, which will be presented and discussed in detail in the discussions (section 8.5.2) and which specifically captures impact transfer from the material to the printing process.

8.3.1.2. *Functional unit*

The functional unit (FU) is another crucial methodological element in LCA. Indeed, the essence of the LCA method is the comparison of products or systems in order to arbitrate between several solutions. The FU enables this comparison on a common basis and is defined as a unit for measuring the function of a system in terms of services rendered.

To better understand the methodological importance of precisely defining the FU, let us compare the impacts of two types of concrete: high-performance concrete (HPC with compressive strength class C90/120) and conventional concrete (C25/30) (see Figure 8.4).

The impact of HPC concrete is far greater than that of conventional concrete, since it contains far more cement (e.g. the quantity can go from 250 kg cement/m^3 of conventional concrete to 500 kg cement/m^3 of HPC). However, it also offers higher compressive strength, potentially reducing the quantity of materials used for the same structural span. Thus, by normalizing the impacts of the two types of concrete per unit volume (kg CO_2 eq/m^3), HPC concrete proves to have a greater impact. On the other hand, when normalizing the impacts per unit of compressive strength (kg CO_2 eq/ MPa), HPC concrete performs better and has less impact (see Agustí-Juan and Habert 2017). It is therefore important to note that an LCA is carried out for a given use.

1 See: https://frq.gouv.qc.ca/histoire-et-rapport/integrer-le-sable-dans-lanalyse-du-cycle-de-vie/.

8.3.2. *Impact model, system boundaries and FU for the 3D concrete printing process*

Unlike traditional construction, where the impact of the installation process is negligible (~1–2%) in the climate change category, the impact of a printed element will be made up of two types of impact:

– *Imat*, the volumetric impact of the material in kg CO_2 eq/m^3;

– *Iproc*, the hourly impact of the process in kg CO_2 eq/h.

Iproc, in turn, consists of:

– *the operational impact*, linked to the printer's electricity consumption and thus defined by the environmental profile of the local/used electricity mix;

– *the intrinsic impact*, resulting from all the impacts linked to the production, transport and installation of the printing system and its components.

Thus, the environmental impact of a 3D-printed element is defined according to the following model:

$$I_{3DCP} = Imat * V + Iproc * t \qquad [8.1]$$

where V is the volume of material and t is the printing time.

By definition: $V = L*S = 1$, and the filament length L is inversely proportional to the print resolution expressed in cross-section S $(S = 1/L)$. So, given that the impact of the process (*Iproc*) is a function of the printing time (t), which in turn is a function of the robot speed (vr) and the distance to be covered (i.e. filament length L $(t=L/vr)$), it is the printing resolution (expressed in cross-section $S(L = V/S= 1/S)$) that will drive, to first order, the impact of the process (*Iproc*).

The impact of the material (*Imat*) and the impact of the process (*Iproc*) are linked by the printing throughput, which defines the quantity of material used per unit of time (m^3/t).

To capture this relationship within the FU, we propose to carry out the analysis at the scale of the elementary brick of the printing process: a printed filament with a total volume V of 1 m^3, cross-section S and length L (see Figure 8.4).

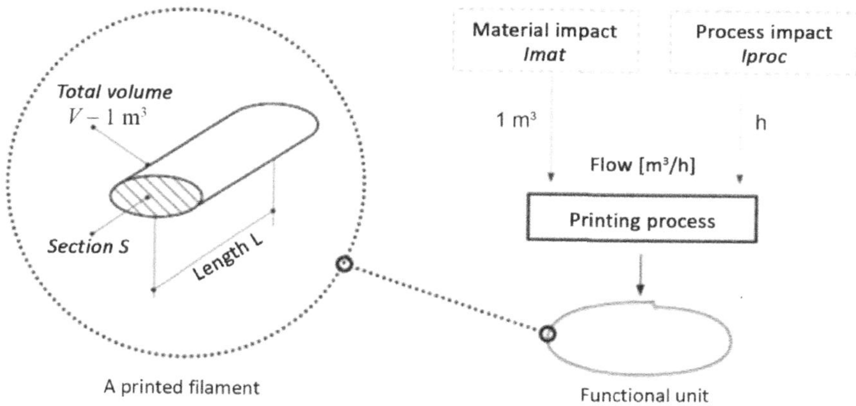

Figure 8.4. *Impact model and functional unit diagram (Kuzmenko 2021)*

This reasoning, based on the relationship between the elementary parameters of the printing process, is universal and will be valid for all of the types of digital concrete system described above. In other words, for all *process-material* pairs, it is possible to characterize the impact of the *Iproc* process and then the impact of the *Imat* material, and thus to evaluate the share of each within 1 m^3 implemented or a manufactured element. Here, we have quantified these values for the following process-material pairing:

– printing system based on an ABB IRB 8700 industrial robot with six axes of freedom, shown in Figure 8.2, *Iproc* = between 2 and 5 kg CO_2 eq/h;

– high-performance printable concrete, *Imat* = between 300 and 500 kg CO_2 eq/m^3.

In this chapter, we concentrate on the impact category of climate change. Data for other impact categories, as well as the inventory used, measurement protocol and calculation methods, can be found in our previous publications (Kuzmenko 2021; Kuzmenko et al. 2022; Roux et al. 2022).

8.3.3. *Impact model and FU for the constructive system 3D concrete printing process*

At the scale of the printed concrete construction system, the impact pattern remains essentially the same as at the process scale (see equation [8.1]). Added to this are the impacts of additional materials introduced into the system for thermal insulation, printing support and reinforcement.

The reference flow used is 1 m^2 of wall. The characteristics of the FU then vary according to the conventional equivalent chosen.

For the comparative study in the masonry category, the target performances of the FU were set according to Eurocode 6 for mechanical resistance, and according to current environmental regulations RT2020 for thermal resistance 6.6 $m^2K \cdot W^{-1}$.

For the comparative study in the reinforced concrete category, target performances for the FU were set according to Eurocode 2 in terms of mechanical resistance and also according to RE2020 in terms of thermal resistance. Structural pre-dimensioning was carried out on the wall sample, 3.3 m high and 0.3 m wide, with clamped lower supports, applying vertical loads of 217 kN and horizontal loads of 6.75 kN in the support plane. Insulation was dimensioned at 5.12 $m^2K \cdot W^{-1}$.

Tables 8.1 and 8.2 summarize the quantities of materials used for the FU of the lattice wall with its mechanical and structural performance as a masonry and reinforced concrete concept.

Inventory	Parameters	Details/source/references
Concrete	60 kg	(Kuzmenko 2021)
3D printing cell	10 minutes	(Kuzmenko 2021)
Insulation	100 kg	Cement mortar | market for cement mortar | fr
Electricity	0.8 kWh	Market for electricity, medium voltage | fr

Table 8.1. *Inventory for the LCA study of the masonry lattice wall*

Inventory	Parameters	Details/source/EcoInvent references
Concrete	60 kg	(Kuzmenko 2021)
3D printing cell	10 minutes	(Kuzmenko 2021)
Insulation	100 kg	Cement mortar | market for cement mortar | fr
Electricity	0.8 kWh	Market for electricity, medium voltage | fr
Fiber	1.4 kg	Glass fiber production | glass fibre | cut-off | rer

Table 8.2. *Inventory for the LCA study of the reinforced concrete lattice wall*

8.4. Results

8.4.1. *Environmental impact on the scale of the printed material*

Figure 8.5 illustrates the impact of 1 m^3 of printed concrete and the variation of the latter as a function of the printing resolution expressed in cross-section *S*. These results show that the cross-sectional area is inversely proportional to the distance covered by the print head and the robot, and that the proportion of the overall impact and that of the *Iproc* process increase as the cross-sectional area decreases. In concrete terms, for the given process–material pair implemented with a filament cross-section of 0.6 cm^2, the impact of 1 m^3 of printed concrete is double that of concrete cast in a formwork: 600 versus 300 kg CO_2 eq/m^3.

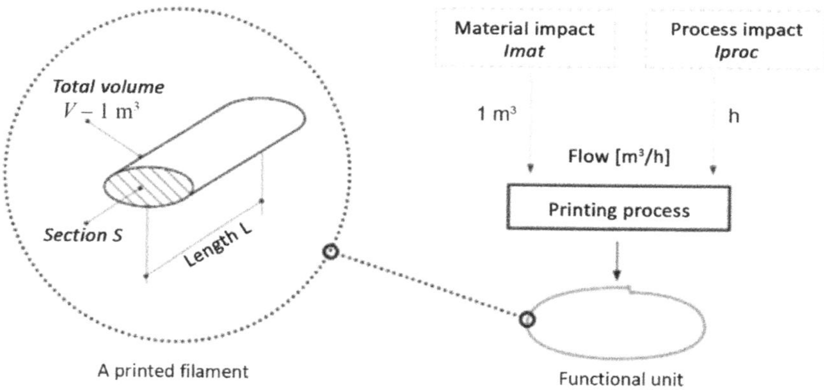

Figure 8.5. *Variation in environmental impact of 1 m^3 printed as a function of print section/resolution (Kuzmenko 2021)*

At this scale of analysis, a coarse printing resolution seems to be a way of reducing the impact of both the printing process and the final element. However, at the scale of the construction system and the industry, the environmental objective will be to reduce the volume of concrete deposited. This implies a strategy of putting material where it is structurally necessary, and therefore of using a thin section for the production of optimized and therefore geometrically complex elements (see Figure 8.2). In other words, the ability to optimize a given geometry will be an increasing function of the printing resolution (see Figure 8.6). There is therefore an optimum printing resolution defined for a given process-material pairing, or rather "application" (more or less optimizable)/"material" (with a given carbon footprint).

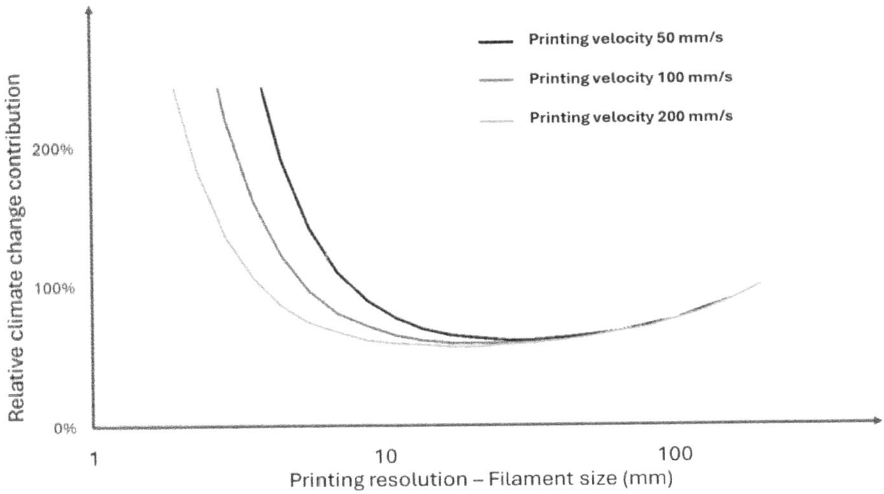

Figure 8.6. *The relative contribution to the carbon footprint studied as a function of printhead resolution for three different speeds (adapted from Kuzmenko et al. (2022))*

Printed concrete therefore inevitably has a greater impact than its cast equivalent. However, at the scale of the construction system, the use of 3D printing technology should nevertheless enable the creation of structures that consume sufficiently less material to offset the additional impact of the printing process, and thus result in an element that performs better overall. This issue is addressed in the next section.

8.4.2. *Environmental impact of the building system*

8.4.2.1. *Masonry*

Figure 8.7 presents a comparison of the environmental impacts of the lattice wall with its conventional masonry analogues and shows that the impact of the space lattice wall is comparable to that of existing breeze-block solutions, without any substantial improvement. In concrete terms, the space lattice system even proves to have a greater impact in the categories of climate change, eutrophication and water pollution. Impacts are broadly similar for the acidification and eutrophication indicator. However, in the other categories, notably air pollution, depletion of abiotic resources and the ozone layer, and photochemical ozone formation, the spatial lattice significantly outperforms its conventional analogues. This is mainly due to the absence of glass wool insulation, usually used in traditional building systems, replaced by cementitious foam.

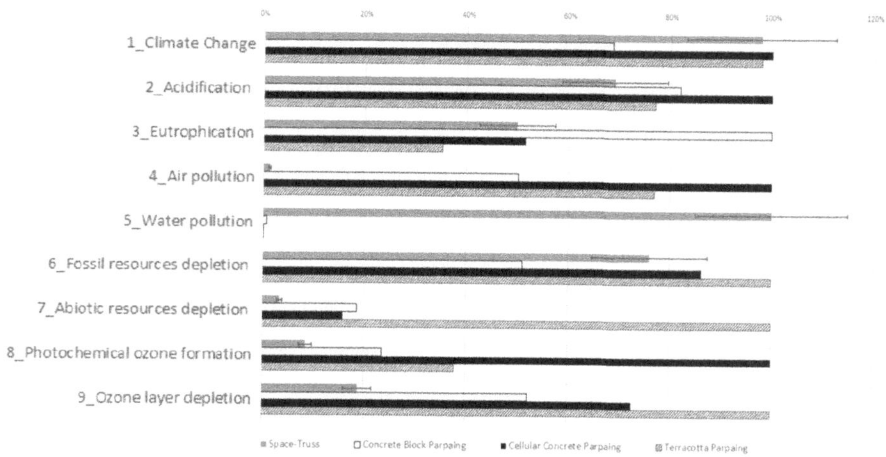

Figure 8.7. *Comparison of the environmental impact of space lattice walls and other masonry systems (Kuzmenko 2021)*

Generally speaking, the comparative study reveals a systematic transfer of impact and pollution within different environmental categories, when comparing one building system to another. In other words, it is hardly possible to say absolutely which system has the most optimal environmental impact. However, focusing specifically on the climate change indicator, a general conclusion can be drawn that, despite significant material savings, the space lattice system does not provide an environmental advantage. Indeed, the structure uses fewer construction materials (minerals in particular), but this reduction is offset by the impact associated with the process and implementation of these materials. In other words, this proves the study's initial hypothesis that the quantity of materials is not an appropriate metric for the environmental optimization of structures.

8.4.2.2. *Reinforced concrete*

Figure 8.8 shows a relative comparison of the environmental impacts of the two structural systems, space lattice and reinforced concrete, versus the reinforced concrete system. The results show that the space lattice reinforced with continuous fiber significantly outperforms the traditional reinforced concrete wall in all environmental categories, with the sole exception of water pollution, where the space lattice's impact is around 30 times higher. The latter stems from the impact of the printing process, notably the nuclear energy used in the French industrial context. Otherwise, the impact of the space lattice wall is half that of its conventional analogue in the categories of climate change, acidification, eutrophication, fossil resources and ozone depletion. In categories such as air pollution, abiotic resource depletion and photochemical ozone formation, the impact of the space lattice wall is less than 5% of that of reinforced concrete.

Overall, even if the current comparison is with pre-dimensioning for space lattice, the scale of the potential improvement in all environmental categories shows that it may be a promising alternative to reinforced concrete.

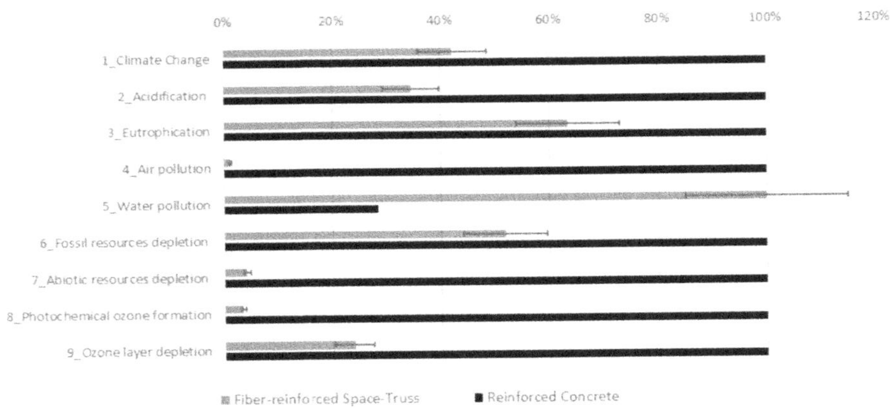

Figure 8.8. *Comparison between space lattice and reinforced concrete walls (Kuzmenko 2021)*

8.5. Discussions and prospects

8.5.1. *Alternative materials for printing*

In the Introduction, we mentioned another way of reducing the impact of concrete construction, namely, by substituting cement in the concrete mix. This development of various "low-carbon concretes" substituting cement with other materials also exists for the practice of 3D printing. In addition, many academic and industrial players are looking to print in raw earth or clay.

However, in view of the reasoning developed above, it is possible to infer that the impact of 1 m^3 of printed earth or clay-based material will be dominated by that of the process. This is because *Imat* (the impact of the soil or clay-based material) will be close to zero, while *Iproc* (the impact of the process) will remain unchanged and may even increase as a result of changes in the material's rheological properties, notably viscosity and shear yield stress.

What is more, considering the low mechanical strength of earth- and clay-based materials, as well as "low-carbon concrete" mixes, few material gains are conceivable on the scale of the constructive system. Indeed, the mechanical compressive strength of earth or clay is only around 2 MPa, while it does not exceed 30 MPa for low-carbon concrete. These values are well below the compressive strength of high-performance concrete (HPC), which can reach up to 120 MPa. So, despite its very high volumetric carbon footprint, we assume that the use of HPC concrete has a far greater potential for structural optimization than its alternative "low-carbon" materials, whether earth- or clay-based, when combined with a 3D-printed construction process, rather than a traditional one.

8.5.2. *Impact transfer*

Finally, we come to the question of the transfer of environmental impact, particularly that of the global warming category considered here, from the material to the process. To assess this aspect, which is central to our reasoning, we introduce the *Kf* factor, developed in greater detail in our previous publications (Kuzmenko et al. 2022). In short, *Kf* is a dimensionless factor that allows us to assess the level of impact transfer from the *Imat* material to the *Iproc* process, and is simply calculated as a ratio between the two. Since *Iproc* decreases with printing resolution (while *Imat* remains unchanged; see Figure 8.6), *Kf* is therefore strictly increasing with printing resolution.

In addition to visualizing the distribution of impact within a given process–material pair, *Kf* can be used to classify, or even steer, suitable and environmentally sustainable industrial applications for the technology under consideration. Let us take a closer look at the three possible *Kf* values:

1) *Kf* =/< 5 %: this means that the printing process observed is neutral in terms of impact, just like traditional processes. This will be the case for 3D printing systems operating with low printing resolution (large cross-section filament), which offers little potential for optimization and impact reduction.

2) *Kf* =/< 50 %: the observed process is far from neutral in terms of impact. Nevertheless, this suggests that, at the scale of a building system, the carbon footprint of the process implemented can be offset by structural material savings, by adopting an optimal and rational material deposition strategy illustrated in Figures 8.1, 8.3 and 8.4.

3) *Kf* > 50 %: this means that the overall impact of the printed part is dominated by the process, and therefore only “niche”, low-volume markets such as the space industry or nuclear waste management may be concerned. This scenario also includes the practice of 3D printing with earth- or clay-based materials, where much of the impact is transferred from the material to the process. Given the sheer volume of the construction market, the practice will no doubt be unsustainable.

With such a *Kf*, an industrial 3D printing application should only be considered when other priorities dominate environmental issues, such as the need for extreme precision, safety, etc., and when the expected parts fall within low-volume markets such as those mentioned above.

8.5.3. *Improving the quality and quantity of available inventory data*

Life cycle analysis relies on the use of databases such as ecoinvent[2], representing emissions of pollutants into the air, water and soil, as well as resource consumption for thousands of production processes (energy production, transport, materials manufacturing, etc.). In the case of 3D concrete printing, our previous work (Roux et al. 2022) highlighted the need for more reliable, diverse and recent inventories in a number of areas. These include data on concrete admixtures (accelerators, plasticizers, etc.). They are generally present in very small quantities in conventional concretes, are sometimes neglected in LCA studies and very few data are available to account for them. Not all families of compounds are represented in environmental databases, whereas the quantity of these products (and their contribution to impact) becomes significant in the case of very high-performance concretes used in 3D printing.

Next, the electrical and electronic components of the printing process (robot, sensor and control cabinet) should be studied in greater detail, with regard to their reliability (service life), composition (use of critical materials) and end-of-life recycling possibilities.

2 See: https://ecoinvent.org/.

8.6. Conclusion

The current development of 3D concrete printing technology is mainly based on its potential to increase the productivity and formal freedom of concrete construction, as well as to reduce the industry's costs and environmental impact. The latter, based on the ability to reduce the amount of material used in construction elements via rational material deposition, is studied in this work using the LCA method applied to an extrusion printing technology.

First, the environmental impact model establishes a generic evaluation framework that can be applied to all extrusion printing processes, whatever the material used. A specific separation between the impact of the material and that of the printing process is maintained throughout the assessment, which is carried out here at the scale of the process and then of the constructive system (the space lattice wall system). The latter is then compared with its conventional analogues in the masonry category, with no improvement in the balance sheet, and then in the reinforced concrete category, showing great potential for reducing the industry's impact.

The results show that, despite the substantial material savings possible with the 3D printing process, environmental impact savings are not guaranteed on the scale of the constructive system, particularly due to the impact linked to the process.

We therefore propose to use the Kf coefficient, which captures the level of impact transfer from the material to the process for a given technology. This enables us to design the most relevant industrial application from an environmental point of view.

8.7. Acknowledgments

The authors would like to thank Nicolas Roussel, Nicolas Ducoulombier and Romain Duballet, with whom most of the work cited in this chapter was conceived and carried out.

8.8. References

Agustí-Juan, I. and Habert, G. (2017). Environmental design guidelines for digital fabrication. *J. Clean. Prod.*, 142, 2780–2791. doi: 10.1016/j.jclepro.2016.10.190.

Buswell, R.A., Leal de Silva, W.R., Jones, S.Z., Dirrenberger, J. (2018). 3D printing using concrete extrusion: A roadmap for research. *Cem. Concr. Res.*, 112, 37–49. doi: 10.1016/j.cemconres.2018.05.006.

Buswell, R.A., Leal da Silva, W.R., Bos, F., Schipper, R., Lowke, D., Hack, N., Kloft, H., Mechtcherine, V., Wangler, T.P., Roussel, N. (2020). A process classification framework for defining and describing digital fabrication with concrete. *Cem. Concr. Res.*, 134, 106068. doi: 10.1016/j.cemconres.2020.106068.

Carneau, P., Mesnil, R., Roussel, N., Baverel, O. (2019). An exploration of 3D printing design space inspired by masonry. In *Proceedings of IASS Annual Symposia*, 6, 1–9. International Association for Shell and Spatial Structures (IASS), Barcelona.

Carneau, P., Mesnil, R., Roussel, N., Baverel, O. (2020). Additive manufacturing of cantilever – From masonry to concrete 3D printing. *Autom. Constr.*, 116, 103184. doi: 10.1016/j.autcon.2020.103184.

Demont, L., Ducoulombier, N., Mesnil, R., Caron, J.-F. (2021). Flow-based pultrusion of continuous fibers for cement-based composite material and additive manufacturing: Rheological and technological requirements. *Compos. Struct.*, 262, 113564. doi: 10.1016/j.compstruct.2021.113564.

Duballet, R. (2019). Building systems in robotic extrusion of cementitious materials. PhD Thesis, Université Paris Est, Paris.

Duballet, R., Baverel, O., Dirrenberger, J. (2019). Space truss masonry walls with robotic mortar extrusion. *Structures*, 18, 41–47. doi: 10.1016/j.istruc.2018.11.003.

Ducoulombier, N., Demont, L., Chateau, C., Bornert, M., Caron, J.-F. (2020). Additive manufacturing of anisotropic concrete: A flow-based pultrusion of continuous fibers in a cementitious matrix. *Procedia Manuf.*, 47, 1070–1077. doi: 10.1016/j.promfg.2020.04.117.

Feraille, A. (2016). Du matériau à l'ouvrage : quelques apports méthodologiques relatifs à l'analyse de Cycle de Vie. Thesis, Laboratoire Navier, Ecole des Pont ParisTech.

Flatt, R.J., Roussel, N., Cheeseman, C. R. (2012). Concrete: An eco material that needs to be improved. *J. Eur. Ceram. Soc.*, 32(11), 11. doi: 10.1016/j.jeurceramsoc.2011.11.012.

Gartner, E. (2004). Industrially interesting approaches to "low-CO_2" cements. *Cem. Concr. Res.*, 34(9), 9. doi: 10.1016/j.cemconres.2004.01.021.

Gaudillière, N., Duballet, R., Bouyssou, C., Mallet, A., Roux, P., Zakeri, M., Dirrenberger, J. (2019). Large-scale additive manufacturing of ultra-high-performance concrete of integrated formwork for truss-shaped pillars. In *Robotic Fabrication in Architecture. Art and Design 2018*. Springer, Cham. doi: 10.1007/978-3-319-92294-2_35.

Gosselin, C., Duballet, R., Roux, Ph., Gaudillière, N., Dirrenberger, J., Morel, Ph. (2016). Large-scale 3D printing of ultra-high performance concrete – A new processing route for architects and builders. *Mater. Des.*, 100, 102–109. doi: 10.1016/j.matdes.2016.03.097.

Hong, J., Shen, G.Q., Feng, Y., Lau, W.S., Mao, C. (2015). Greenhouse gas emissions during the construction phase of a building: A case study in China. *J. Clean. Prod.*, 103, 249–259. doi: 10.1016/j.jclepro.2014.11.023.

Khoshnevis, B. (2004). Automated construction by contour crafting – Related robotics and information technologies. *Autom. Constr.*, 13(1), 1. doi: 10.1016/j.autcon.2003.08.012.

Kuzmenko, K. (2021). Environmental performance in construction: Case-study of 3D concrete printing. Modelling and characterization of impacts. Application to engineered structures. Thesis, Ecole des Ponts ParisTech, Paris.

Kuzmenko, K., Ducoulombier, N., Feraille, A., Roussel, N. (2022). Environmental impact of extrusion-based additive manufacturing: Generic model, power measurements and influence of printing resolution. *Cem. Concr. Res.*, 157, 106807. doi: 10.1016/j.cemconres.2022.106807.

Li, H., Deng, Q., Zhang, J., Xia, B., Skitmore, M. (2019). Assessing the life cycle CO2 emissions of reinforced concrete structures: Four cases from China. *J. Clean. Prod.*, 210, 1496–1506. doi: 10.1016/j.jclepro.2018.11.102.

Pegna, J. (1997). Exploratory investigation of solid freeform construction. *Autom. Constr.*, 5(5), 5. doi: 10.1016/S0926-5805(96)00166-5.

Perrot, A., Rangeard, D., Amziane, S., Sonebi, M. (2018). Défis à relever en termes de rhéologie des matériaux cimentaires pour une impression 3D par extrusion/dépôt. *Acad. J. Civ. Eng.*, 36(1), 1. doi: 10.26168/ajce.36.1.43.

Rippmann, M., Liew, A., Van Mele, T., Block, P. (2018). Design, fabrication and testing of discrete 3D sand-printed floor prototypes. *Mater. Today Commun.*, 15, 254–259. doi: 10.1016/j.mtcomm.2018.03.005.

Roux, C., Kuzmenko, K., Roussel, N., Mesnil, R., Feraille, A. (2022). Life cycle assessment of a concrete 3D printing process. *Int. J. Life Cycle Assess.* doi: 10.1007/s11367-022-02111-3.

Ryklová, P., Mančík, Š., Lupíšek, A. (2019). Environmental benefits of timber-concrete prefabricated construction system for apartment buildings – A simplified comparative LCA study. *IOP Conf. Ser. Earth Environ. Sci.*, 290, 012083. doi: 10.1088/1755-1315/290/1/012083.

Salet, T.A.M., Ahmed, Z.Y., Bos, F.P., Laagland, H.L.M. (2018). Design of a 3D printed concrete bridge by testing. *Virtual Phys. Prototyp.*, 13(3), 3. doi: 10.1080/17452759.2018.1476064.

Scrivener, K.L., John, V.M., Gartner, E.M. (2018). Eco-efficient cements: Potential economically viable solutions for a low-CO_2 cement-based materials industry. *Cem. Concr. Res.*, 114, 2–26. doi: 10.1016/j.cemconres.2018.03.015.

Union Nationale des Producteurs de Granulats (2017). Module d'information environnementale de la production de granulats à partir de roches meubles. Conforming aux normes NF EN 15804+A1 and NF EN 15804/CN. Report.

Van Damme, H. (2018). Concrete material science: Past, present, and future innovations. *Cem. Concr. Res.*, 112, 5–24. doi: 10.1016/j.cemconres.2018.05.002.

Wangler, T., Lloret, E., Reiter, L., Hack, N., Gramazio, F., Kohler, M., Bernhard, M., Dillenburger, B., Buchli, J., Roussel, N. et al. (2016). Digital concrete: Opportunities and challenges. *RILEM Tech. Lett.*, 1, 67. doi: 10.21809/rilemtechlett.2016.16.

Wangler, T., Roussel, N., Bos, F.P., Salet, T.A.M., Flatt, R.J. (2019). Digital concrete: A review. *Cem. Concr. Res.*, 123, 105780. doi: 10.1016/j.cemconres.2019.105780.

List of Authors

Sofiane AMZIANE
Institut Pascal
Université Clermont-Auvergne
Clermont-Ferrand
France

Olivier BAVEREL
Laboratoire Navier
École des Ponts
Marne-la-Vallée
France

Victor DE BONO
XtreeE
Rungis
France

Jean-François CARON
Laboratoire Navier
École des Ponts
Marne-la-Vallée
France

Léo DEMONT
Laboratoire Navier
École des Ponts
Marne-la-Vallée
France

Zeinab DIAB
Lamé
Université d'Orléans
France

Duc-Phi DO
Lamé
Université d'Orléans
France

Romain DUBALLET
XtreeE
Rungis
France

Nicolas DUCOULOMBIER
XtreeE
Rungis
France

Adélaïde FERRAILLE
Laboratoire Navier
École des Ponts
Marne-la-Vallée
France

Dashnor HOXHA
Lamé
Université d'Orléans
France

Yohan JACQUET
Technische Universität Berlin
Germany

Fatima Zahraa KACHKOUCH
IRDL
Université de Bretagne Sud
Lorient
France

Kateryna KUZMENKO
Groupe Kardham
Paris
France

Romain MESNIL
Laboratoire Navier
École des Ponts
Marne-la-Vallée
France

Arnaud PERROT
IRDL
Université de Bretagne Sud
Lorient
France

Alexandre PIERRE
CY Cergy Paris Université
France

Sébastien RÉMOND
Lamé
Université d'Orléans
France

Charlotte ROUX
École des Mines de Paris
France

Index

I, L, M

N, P, Q

R, S

T, V

Printed and bound by CPI Group (UK) Ltd, Croydon, CR0 4YY
16/09/2026
14963196-0002